Alexander Stoppa

Chemical Speciation in Mixtures of Ionic Liquids and Polar Compounds

Alexander Stoppa

Chemical Speciation in Mixtures of Ionic Liquids and Polar Compounds

Studies into the Structure and Dynamics

Südwestdeutscher Verlag für Hochschulschriften

Imprint
Any brand names and product names mentioned in this book are subject to trademark, brand or patent protection and are trademarks or registered trademarks of their respective holders. The use of brand names, product names, common names, trade names, product descriptions etc. even without a particular marking in this work is in no way to be construed to mean that such names may be regarded as unrestricted in respect of trademark and brand protection legislation and could thus be used by anyone.

Publisher:
Südwestdeutscher Verlag für Hochschulschriften
is a trademark of
Dodo Books Indian Ocean Ltd., member of the OmniScriptum S.R.L Publishing group
str. A.Russo 15, of. 61, Chisinau-2068, Republic of Moldova Europe
Printed at: see last page
ISBN: 978-3-8381-2042-3

Zugl. / Approved by: Regensburg, Universität Regensburg, Dissertation, 2010

Copyright © Alexander Stoppa
Copyright © 2010 Dodo Books Indian Ocean Ltd., member of the OmniScriptum S.R.L Publishing group

Contents

Introduction 1

1 Theoretical background 5
   1.1 Basics of electrodynamics 5
      1.1.1 Maxwell and constitutive equations 5
      1.1.2 The electric displacement field 6
      1.1.3 Wave equations 7
   1.2 Dielectric relaxation 8
      1.2.1 Polarization 8
      1.2.2 Response functions of the orientational polarization 9
   1.3 Empirical description of dielectric relaxation 10
      1.3.1 Debye equation 10
      1.3.2 Distribution functions 11
      1.3.3 Damped harmonic oscillator 12
      1.3.4 Combination of models 13
      1.3.5 Data processing 13
   1.4 Microscopic models of dielectric relaxation 14
      1.4.1 Onsager equation 14
      1.4.2 Cavell equation 16
      1.4.3 Debye model of rotational diffusion 17
      1.4.4 Microscopic and macroscopic relaxation times ... 18
   1.5 Ion pair relaxation 19
      1.5.1 Amplitudes 19
      1.5.2 Kinetics 20
   1.6 Temperature dependence of relaxation times 21

2 Experimental 23
- 2.1 Materials 23
  - 2.1.1 Solvents 23
  - 2.1.2 Ionic liquids 24
  - 2.1.3 Sample preparation 27
- 2.2 Measurement of dielectric properties 28
  - 2.2.1 Frequency-domain reflectometry 28
  - 2.2.2 Interferometry 30
  - 2.2.3 THz time-domain spectroscopy 34
- 2.3 Supplementary measurements 39
  - 2.3.1 Density 39
  - 2.3.2 Conductivity 39
  - 2.3.3 Viscosity 40
  - 2.3.4 Refractive indices 41
- 2.4 Raman spectroscopy 41

3 Neat Components 43
- 3.1 Ionic liquids 43
- 3.2 Acetonitrile 46
- 3.3 Methanol 51

4 IL + IL mixtures 53
- 4.1 Fit model 53
- 4.2 Results 54
- 4.3 Discussion 57

5 IL + polar solvent mixtures 65
- 5.1 Supplementary measurements 65
- 5.2 Dielectric properties 77
  - 5.2.1 IL + acetonitrile mixtures 77
  - 5.2.2 IL + methanol mixtures 106
- 5.3 Raman spectroscopy of IL + acetonitrile mixtures 113

Summary and conclusions 119

Appendix 123
- A.1 Physico-chemical data for [emim][EtSO$_4$] + acetonitrile mixtures 123

Vorwort

Die vorliegende Dissertation entstand in der Zeit von November 2006 bis Juni 2010 am Institut für Physikalische und Theoretische Chemie der naturwissenschaftlichen Fakultät IV—Chemie und Pharmazie—der Universität Regensburg.

An erster Stelle bedanke ich mich bei Herrn Prof. Dr. Richard Buchner für die Erteilung des Themas. Insbesondere seine stete Bereitschaft, bei allen erdenklichen Fragen und Problemen mit wertvollen Ratschlägen zur Seite zu stehen, hat mich persönlich beeindruckt sowie wesentlich zum Fortgang dieser Dissertation beigetragen.

Weiterhin danke ich dem Leiter des Lehrstuhls, Herrn Prof. Dr. Werner Kunz, für seine großzügige Unterstützung. Ferner möchte ich folgende Personen und Institutionen würdigend hervorheben:

- I appreciate the personally enriching support by Prof. Dr. Glenn Hefter, Murdoch University, Perth, Australia, during my stay from October until December 2007 in his group. His interest in my work in the last years and the proofreading of the English text of the present thesis are invaluable for me. The hospitality of the Hefter family was very impressive and will be unforgettable for me. Furthermore, I would like to thank all members of the group for welcoming me so warmly, particularly Dr. Chandrika Akilan and Dr. Zoltán Paksi.

- Dr. Markus Walther und Dr. Andreas Thoman, Institut für molekulare und optische Physik (Prof. Dr. Helm), Albert-Ludwigs-Universität Freiburg, waren wertvolle Kooperationspartner, die ich fachlich sowie persönlich zu schätzen lernte.

- Prof. Dr. Marija Bešter Rogač, Faculty of Chemistry and Chemical Technology, University of Ljubljana, sei für die fruchtbare Zusammenarbeit auf dem Gebiet der Leitfähigkeitsmessungen gedankt.

- Prof. Dr. Augustinus Asenbaum und Dr. Christian Pruner, Institut für Physik und Biophysik, Universität Salzburg, haben mich überaus freundlich in Salzburg aufgenommen und mich mit wertvollen Hinweisen zu Raman Messungen versehen.

- Cornelia Schöggl-Wagner und Thomas Feischl, Anton Paar Gmbh, Graz, danke ich die Durchführung von Viskositäts- und Dichtemessungen.

- Die Deutsche Forschungsgemeinschaft hat diese Arbeit im Rahmen des Schwerpunktprogrammes 'Ionische Flüssigkeiten' (DFG SPP 1191) finanziert und zahlreiche Möglichkeiten zu fachlichem Austausch sowie zum Aufbau weiterer Kooperationen eröffnet.

- Mein Dank geht ferner an die Freunde der Universität Regensburg e.V., die mir durch finanzielle Unterstützung die Teilnahme an mehreren internationalen Tagungen ermöglichten.

Allen Mitarbeitern und Kollegen des Lehrstuhls, insbesondere des Arbeitskreises Mikrowellen (Dr. Simon Schrödle, Dr. Wolfgang Wachter, Dr. Johannes Hunger, Saadia Shaukat und Hafiz Abd Ur Rahman) danke ich für die freundschaftliche Atmosphäre. Den Mitgliedern der mittäglichen Kaffeerunde rechne ich ihre aufmunternden Worte und Erzählungen aus vergangenen Zeiten hoch an.

Nicht zuletzt bedanke ich mich bei allen Mitarbeitern der Werkstätten für die schnelle und ordentliche Erledigung der Aufträge.

Constants, symbols and acronyms

Constants

elementary charge	e_0	$= 1.60217739 \cdot 10^{-19}\,\text{C}$
permittivity of free space	ε_0	$= 8.854187816 \cdot 10^{-12}\,\text{C}^2(\text{Jm})^{-1}$
Avogadro's constant	N_A	$= 6.0221367 \cdot 10^{23}\,\text{mol}^{-1}$
speed of light	c_0	$= 2.99792458 \cdot 10^8\,\text{m}\,\text{s}^{-1}$
Boltzmann's constant	k_B	$= 1.380658 \cdot 10^{-23}\,\text{J}\,\text{K}^{-1}$
permeability of free space	μ_0	$= 4\pi \cdot 10^{-7}\,(\text{Js})^2(\text{C}^2\text{m})^{-1}$
Planck's constant	h	$= 6.6260755 \cdot 10^{-34}\,\text{Js}$

Symbols

$\vec{B}$	magnetic induction [Vs m^{-2}]		$\vec{D}$	electric induction [C m^{-2}]
$\vec{E}$	electric field strength [V m^{-1}]		$\vec{H}$	magnetic field strength [A m^{-1}]
$\vec{P}$	polarization [C m^{-2}]		μ	dipole moment [C m]
$\hat{\varepsilon}$	complex dielectric permittivity		ε'	real part of $\hat{\varepsilon}$
ε''	imaginary part of $\hat{\varepsilon}$		ε	$\lim_{\nu \to 0}(\varepsilon')$
ε_∞	$\lim_{\nu \to \infty}(\varepsilon')$		τ	relaxation time [s]
T	thermodynamic temperature [K]		θ	temperature [°C]
ν	frequency [s^{-1}]		ω	angular frequency [s^{-1}]

Acronyms

AN	acetonitrile	BN	benzonitrile
1-BuOH	1-butanol	DCM	dichloromethane
DMA	N,N-dimethylacetamide	DMSO	dimethylsulfoxide
MeOH	methanol	PC	propylene carbonate
W	water	(RT)IL	(room-temperature) ionic liquid
DR	dielectric relaxation	FIR	far-infrared
IF	interferometry	TDR	time-domain reflectometry
TDS	time-domain spectroscopy	VNA	vector network analyzer
D	Debye	CC	Cole-Cole
CD	Cole-Davidson	DHO	damped harmonic oscillator

Introduction

Basic aspects

Being defined as salts that melt below 100 °C, ionic liquids (ILs) are regarded as an exciting class of versatile materials. Particularly room-temperature ionic liquids (RTILs) have attracted much current interest, because they combine a number of outstanding properties. Most importantly, they have a wide liquid range with melting points around ambient temperature, high thermal and electrochemical stabilities, low volatility and flammability, as well as the option of tuning various physical and chemical properties by a straightforward change of cations or anions,[1] which has led to the term 'designer solvents'. In addition, the available property range of ILs can be enlarged even further by using binary IL + IL mixtures as this allows continuous and simultaneous adjustment of several key properties, such as viscosity, conductivity and polarity. To date, industrial-scale applications of IL + IL mixtures are scarce but, for instance, they seem to be promising replacements for organic solvents in the production of dye-sensitized solar cells.[2,3] One reason for their limited use so far is almost certainly the lack of data for relevant solvent properties. Only a small number of papers dealing with volumetric and transport properties[4–7] has been published so far. Some studies using solvatochromic probes[8] and optical heterodyne-detected Raman-induced Kerr effect (OHD-RIKE) spectroscopy have also been reported.[9,10] These last investigations showed that for some mixtures the OHD-RIKE spectra, covering $0.1 - 6$ THz, were mole-fraction-weighted averages of the pure-component spectra. For other mixtures such additivity did not hold, suggesting different mixing states.[9,10]

Neat ILs have been used in various applications, for example as electron or proton conductors in battery research, as electrolytes in electrochemistry, and as solvents in catalytic and extraction processes or in chemical synthesis.[1,11–15] ILs can further be applied in biochemistry for the stabilization of enzymes[16] or for dissolution of (self-assembling) carbohydrates.[17] Nevertheless, despite their outstanding properties, one should always keep in mind the effects of a specific property when choosing an IL for an application. For example, the widely recognized dissolution capacity of ILs for various compounds makes their use in supercritical CO_2 extraction processes possible,[18] but on the other hand, the removal of highly soluble compounds for purification of

ILs becomes a demanding task.

For technological purposes ILs will rarely be employed in neat form: almost invariably they will be diluted either by reactants and products or by the presence of a co-solvent, deliberately added to optimize the physical and chemical characteristics of the IL. It is essential to be aware of the effects of added compounds, either impurities or co-solvents, on the physico-chemical properties of ILs. The initial study dealing with these effects was reported by Seddon et al.[19] Since that time a growing number of studies has appeared reporting physico-chemical data of IL + solvent mixtures, but the coverage and quality of the available data are not satisfactory. In particular reliable measurements of various transport properties are rarely available (see refs. 20–22 and literature cited therein). Although knowledge of these physico-chemical properties is important for technological applications such quantities do not appear to yield any structural and dynamical information.

Concerning the speciation in IL + solvent mixtures, they have been experimentally studied by solvatochromic probes,[23,24] UV-Vis and IR spectroscopies,[25,26] NMR-spectroscopy,[27,28] mass spectrometry[29] and Brillouin light scattering.[30] Of special relevance for the present work, the main conclusions may be highlighted as follows: 1. ILs keep their nanostructured organization even upon dilution with polar solvents, 2. discrete ion pairs formed by the IL ions are present in low-permittivity solvents, and 3. IL cations are solvated by solvent molecules.

This experimental view is supported by a few molecular dynamics (MD) simulations. Wu et al.[31] studied IL + acetonitrile (AN) mixtures and concluded, that cation-anion and AN-AN interactions are enhanced after mixing, leading to negative deviations from ideal solutions. The picture presented by Pádua et al.[32,33] is more depictive: AN and methanol (MeOH) interact with both the charged (polar part of the cations together with anions) and the non-polar domains (non-polar parts of the cations) present in ILs.[32,33]

Being an useful tool for the investigation of the structure and dynamics of electrolyte solutions in general[34] and neat ionic liquids in particular,[35–41] dielectric relaxation (DR) spectroscopy[42] is a promising technique. It probes the fluctuations of permanent dipoles in response to the application of an oscillating electromagnetic field in the microwave (GHz) region. DR spectroscopy is therefore sensitive to reorientational and cooperative motions of dipolar species on the pico- to nanosecond timescale.[34] It is especially sensitive to the presence of ion pairs,[34,43,44] which have often been invoked to explain various IL properties, such as their relatively low conductance and high viscosity,[45] although the existence of such species has been excluded by DR studies[46] and MD simlulations[47] of neat imidazolium-based ILs.

However, the situation is rather unsatisfactory regarding studies into the structure, particularly speciation, and dynamics of IL + IL and IL + solvent mixtures. The number of systematic

INTRODUCTION

studies with respect to the frequency range covered and the number of compositions studied is rather limited.[9,10,37,48]

To allow detailed and quantitative analysis, measurements of the DR spectra over a sufficiently broad frequency range ($0.2 \leq \nu/\text{GHz} \leq 20$ or ≤ 89) and supplementary measurements (conductivities and densities) of various binary mixtures have been performed and the results are presented in this thesis. The scope of these experiments is twofold. Firstly, reliable values of selected physico-chemical properties (conductivities, densities, molar conductivities and excess molar volumes) of various mixtures are provided. Then, the structure and dynamics of the species present in the mixtures are analyzed in more detail, yielding information on ion-ion and ion-solvent interactions and the transition from molten-salt like to electrolyte-solution like behavior.

Systems investigated

The first part of this PhD thesis will provide an overview of DR studies of the neat components used in the present work (ILs, AN and MeOH) and their most important results. The knowledge of the mechanism governing their DR spectra will be essential for the following studies.

As a model system for binary IL + IL mixtures, consisting of 1-ethyl-3-methylimidazolium tetrafluoroborate ([emim][BF$_4$]) and its dicyanamide ([emim][DCA]) was chosen, as these ILs are readily available in high purity and fully miscible over the whole composition range. Moreover, their dynamics are fast (compared to other ILs) on the DR timescale and their anion dipole moments are zero (BF$_4^-$) or small (DCA$^-$).[49] These properties simplify the analysis of the DR spectra.

The main part of the present PhD thesis deals with binary mixtures of various ILs and AN covering, wherever possible, the whole composition range.

AN is a dipolar, aprotic (protophobic) solvent (gas phase dipole moment $\mu = 3.96\,\text{D}$),[50] which has been widely employed, for example, in the hydrometallurgical processing of Cu,[51] in battery applications,[52] or as a popular solvent in liquid chromatography.[53] It has a convenient liquid range, a relatively low viscosity ($\eta = 0.3413\,\text{mPa\,s}$ at $25\,°\text{C}$),[54] a reasonably high dielectric constant ($\varepsilon = 35.96$ at $25\,°\text{C}$),[54] and the ability to dissolve a wide range of organic and inorganic compounds.[52,54,55] AN was chosen, as it is known to be fully miscible at ambient temperatures with many alkylimidazolium-based ILs.[22,24,30,56] The ILs studied were the following: the tetrafluoroborate salts of 1-ethyl-3-methylimidazolium ([emim][BF$_4$]), 1-butyl-3-methylimidazolium ([bmim][BF$_4$]), and 1-hexyl-3-methylimidazolium ([hmim][BF$_4$]), the chloride ([bmim][Cl], dilute mixtures), hexafluorophosphate ([bmim][PF$_6$]) and dicyanamide salts ([bmim][DCA]) of 1-butyl-3-methylimidazolium, as well as 1-hexyl-3-methylimidazolium bis-[(trifluoromethyl)sulfonyl]imide ([hmim][NTf$_2$]) and 1-ethyl-3-methylimidazolium ethylsulfate ([emim][EtSO$_4$]). Imidazolium-based ILs were chosen as they are the most intensively studied ILs[15] and they are readily prepared or commercially available in reasonable purity.[22] The effect of the cation and/or the anion on the properties of IL + AN mixtures was studied by variation of the chain-length of the 1-hydrocarbon substituent on the imidazolium ring and/or the anion itself. Additionally, the system [bmim][BF$_4$] + MeOH was chosen for the investigation of different types of interactions exerted by amphiprotic hydrophilic (MeOH, $\eta = 0.5438$ mPa s, $\varepsilon = 32.63$ at 25 °C)[57] and aprotic protophobic solvents (AN) with the IL. Complementary to previously published DR studies of IL + dichloromethane mixtures,[37,48] these experiments were performed to yield more detailed insights into the speciation in IL + polar solvent mixtures.

Chapter 1

Theoretical background

1.1 Basics of electrodynamics

1.1.1 Maxwell and constitutive equations

Maxwell's equations,[58,59] a set of four partial differential equations,

$$\vec{\text{rot}}\, \vec{H} = \vec{j} + \frac{\partial}{\partial t}\vec{D} \tag{1.1}$$

$$\vec{\text{rot}}\, \vec{E} = -\frac{\partial}{\partial t}\vec{B} \tag{1.2}$$

$$\text{div}\, \vec{D} = \rho_{\text{el}} \tag{1.3}$$

$$\text{div}\, \vec{B} = 0, \tag{1.4}$$

relate electric, $\vec{E}$, and magnetic fields, $\vec{H}$, to their sources, charge density, ρ_{el}, and current density, $\vec{j}$. Gauss' law for magnetic fields (Eq. 1.4, magnetic induction $\vec{B}$) expresses the absence of magnetic charges, Gauss' law for electric fields (Eq. 1.3) the production of electric induction (or electric displacement field), $\vec{D}$, by electric charges, Faraday's law of induction (Eq. 1.2) the formation of electric fields by changing magnetic fields and Ampère's circuital law (Eq. 1.1) the generation of magnetic fields by currents.

These four equations, together with the Newton equation,

$$m\frac{\partial^2}{\partial t^2}\vec{r} = q(\vec{E} + \vec{v} \times \vec{B}), \tag{1.5}$$

where q corresponds to a moving charge with velocity $\vec{v}$, define the complete set of laws of classical electromagnetism, which allows a full description of electromagnetic phenomena.

To apply Maxwell's equations to homogenous, nondispersive, isotropic materials at low fields, the constitutive equations,

$$\vec{D} = \varepsilon\varepsilon_0\vec{E} \tag{1.6}$$

$$\vec{j} = \kappa\vec{E} \tag{1.7}$$

$$\vec{H} = \frac{\vec{B}}{\mu\mu_0}, \tag{1.8}$$

where ε_0 and μ_0 are the permittivity and permeability of free space, respectively, are introduced. They connect $\vec{D}$ and $\vec{H}$ fields to $\vec{E}$ and $\vec{B}$ by time- and field strength-independent scalars (material properties): the relative permittivity, ε, electric conductivity, κ, and relative permeability, μ.

1.1.2 The electric displacement field

The constitutive equations are valid for a time-independent field response. For most materials, Eqs. 1.6-1.8 are not simple proportionalities but, rather, are functions of frequency.
The dynamic case may be studied by applying an harmonically oscillating electric field,

$$\vec{E}(t) = \vec{E}_0 \cos(\omega t), \tag{1.9}$$

where $\vec{E}_0$ is the amplitude and ω the angular frequency. When the frequency is sufficiently high (in the order of $1\,\mathrm{MHz}$ to $1\,\mathrm{GHz}$), the motions of the particles in a typical condensed phase cannot follow the field changes and thus most such phases show a phase delay, $\delta(\omega)$, between the electric field and the electric displacement field,

$$\vec{D}(t) = \vec{D}_0 \cos(\omega t - \delta(\omega)). \tag{1.10}$$

In Eq. 1.10, $\vec{D}_0$ is the amplitude of the harmonic oscillation. By using the cosine difference formula and subsequent introduction of

$$\vec{D}_0 \cos(\delta(\omega)) = \varepsilon'(\omega)\varepsilon_0\vec{E}_0 \tag{1.11}$$

$$\vec{D}_0 \sin(\delta(\omega)) = \varepsilon''(\omega)\varepsilon_0\vec{E}_0, \tag{1.12}$$

the electric displacement field is expressed as

$$\vec{D}(t) = \varepsilon'(\omega)\varepsilon_0\vec{E}_0 \cos(\omega t) + \varepsilon''(\omega)\varepsilon_0\vec{E}_0 \sin(\omega t), \tag{1.13}$$

and the phase delay as

$$\tan(\delta(\omega)) = \frac{\varepsilon''(\omega)}{\varepsilon'(\omega)}. \tag{1.14}$$

1.1. BASICS OF ELECTRODYNAMICS

Now, $\vec{D}(t)$ is characterized by a dispersive part (first term in Eq. 1.13), which is in-phase with $\vec{E}(t)$, and a phase-shifted dissipative term (second term in Eq. 1.13). The dielectric dispersion, $\varepsilon'(\omega)$, and dielectric loss, $\varepsilon''(\omega)$ contributions are summarized as the complex permittivity,

$$\hat{\varepsilon}(\omega) = \varepsilon'(\omega) - \mathrm{i}\varepsilon''(\omega). \tag{1.15}$$

By using complex notation, the complex field vectors $\hat{\vec{E}}(t)$ and $\hat{\vec{D}}(t)$ are introduced via

$$\hat{\vec{E}}(t) = \vec{E}_0 \cos(\omega t) + \mathrm{i}\vec{E}_0 \sin(\omega t) = \vec{E}_0 \exp(\mathrm{i}\omega t) \tag{1.16}$$

$$\hat{\vec{D}}(t) = \vec{D}_0 \cos(\omega t - \delta) + \mathrm{i}\vec{D}_0 \sin(\omega t - \delta) = \vec{D}_0 \exp[\mathrm{i}(\omega t - \delta)]. \tag{1.17}$$

Thus, the complex form of the constitutive equations is obtained for the dynamic, i.e. frequency dependent, case as[60]

$$\hat{\vec{D}}(t) = \hat{\varepsilon}(\omega)\varepsilon_0 \hat{\vec{E}}(t) \tag{1.18}$$

$$\hat{\vec{j}}(t) = \hat{\kappa}(\omega)\hat{\vec{E}}(t) \tag{1.19}$$

$$\hat{\vec{B}}(t) = \hat{\mu}(\omega)\mu_0 \hat{\vec{H}}(t) \tag{1.20}$$

with the complex conductivity, $\hat{\kappa}(\omega)$, and the complex relative magnetic permeability, $\hat{\mu}(\omega)$.

1.1.3 Wave equations

Assuming harmonically oscillating fields $\hat{\vec{E}}(t) = \vec{E}_0 \exp(\mathrm{i}\omega t)$ and $\hat{\vec{H}}(t) = \vec{H}_0 \exp(\mathrm{i}\omega t)$, Ampère's law (Eq. 1.1) and Faraday's law (Eq. 1.2) can be converted with the help of the complex constitutive equations into

$$\vec{\mathrm{rot}}\, \vec{H}_0 = (\hat{\kappa}(\omega) + \mathrm{i}\omega\hat{\varepsilon}(\omega)\varepsilon_0)\vec{E}_0 \quad \text{and} \tag{1.21}$$

$$\vec{\mathrm{rot}}\, \vec{E}_0 = -\mathrm{i}\omega\hat{\mu}(\omega)\mu_0 \vec{H}_0. \tag{1.22}$$

Applying the Legendre vectorial identity,

$$\vec{\mathrm{rot}}\, \vec{\mathrm{rot}}\, \vec{H}_0 = \vec{\mathrm{grad}}\,\mathrm{div}\, \vec{H}_0 - \triangle \vec{H}_0 = \vec{\mathrm{grad}}\,(0) - \triangle \vec{H}_0 = -\triangle \vec{H}_0, \tag{1.23}$$

one obtains from combination of Eqs. 1.21 and 1.22 the reduced form of the wave equation of the magnetic field as

$$\triangle \vec{H}_0 + \hat{k}^2 \vec{H}_0 = 0. \tag{1.24}$$

The propagation constant, $\hat{k}$, is defined as

$$\hat{k}^2 = k_0^2 \left(\hat{\mu}(\omega)\hat{\varepsilon}(\omega) + \frac{\hat{\mu}(\omega)\hat{\kappa}(\omega)}{\mathrm{i}\omega\varepsilon_0}\right). \tag{1.25}$$

The propagation constant of free space, k_0, is given by

$$k_0 = \omega\sqrt{\varepsilon_0\mu_0} = \frac{2\pi}{\lambda_0} \quad \text{with} \tag{1.26}$$

$$c_0 = \frac{1}{\sqrt{\varepsilon_0\mu_0}}, \tag{1.27}$$

where c_0 and λ_0 are the speed of light and the wavelength of a monochromatic wave in vacuum, respectively. Accordingly, one obtains a reduced wave equation for electric fields in the case of a source-free medium (div $\vec{E} = 0$) as

$$\triangle \hat{\vec{E}}_0 + \hat{k}^2 \hat{\vec{E}}_0 = 0. \tag{1.28}$$

For nonmagnetizable materials ($\hat{\mu} = 1$), $\hat{k}$ is written as

$$\hat{k}^2 = k_0^2\left(\hat{\varepsilon}(\omega) + \frac{\hat{\kappa}(\omega)}{i\omega\varepsilon_0}\right) \equiv k_0^2 \hat{\eta}(\omega) \tag{1.29}$$

and the generalized complex permittivity, $\hat{\eta}(\omega) = \eta'(\omega) - i\eta''(\omega)$, is defined with its real and imaginary parts,

$$\eta'(\omega) = \varepsilon'(\omega) - \frac{\kappa''(\omega)}{\omega\varepsilon_0} \tag{1.30}$$

$$\eta''(\omega) = \varepsilon''(\omega) + \frac{\kappa'(\omega)}{\omega\varepsilon_0}. \tag{1.31}$$

As only $\hat{\eta}(\omega)$ is experimentally accessible, these equations show that dielectric properties and the conductivity of a system cannot be measured separately. Using the limits of $\hat{\kappa}(\omega)$, i.e. $\lim_{\nu\to 0}\kappa' = \kappa$ and $\lim_{\nu\to 0}\kappa'' = 0$, where κ is the dc conductivity, one can calculate the complex dielectric permittivity from $\hat{\eta}(\omega)$ via

$$\varepsilon'(\omega) = \eta'(\omega) \quad \text{and} \tag{1.32}$$

$$\varepsilon''(\omega) = \eta''(\omega) - \frac{\kappa}{\omega\varepsilon_0}. \tag{1.33}$$

Thus, the frequency-dependent part of $\hat{\kappa}(\omega)$ is subsumed in $\hat{\varepsilon}(\omega)$.

1.2 Dielectric relaxation

1.2.1 Polarization

The electric displacement field can be written as a sum of two contributions,

$$\hat{\vec{D}} = \hat{\varepsilon}\varepsilon_0\hat{\vec{E}} = \varepsilon_0\hat{\vec{E}} + \hat{\vec{P}} \quad \text{with} \tag{1.34}$$

$$\hat{\vec{P}} = (\hat{\varepsilon} - 1)\varepsilon_0\hat{\vec{E}}, \tag{1.35}$$

1.2. DIELECTRIC RELAXATION

where the polarization, $\hat{\vec{P}}$, is a measure of the induced macroscopic dipole moment in the medium due to an applied electric field, whereas $\varepsilon_0 \hat{\vec{E}}$ is the contribution to $\hat{\vec{D}}$, which is also present in vacuum.

Going to the microscopic level, the macroscopic polarization, $\hat{\vec{P}}$, is the sum of orientational, $\hat{\vec{P}}_\mu$, and induced, $\hat{\vec{P}}_\alpha$, polarizations.[60] These are defined as

$$\hat{\vec{P}}_\mu = \sum_k \rho_k \langle \vec{\mu}_k \rangle \quad \text{and} \tag{1.36}$$

$$\hat{\vec{P}}_\alpha = \sum_k \rho_k \alpha_k (\hat{\vec{E}}_i)_k. \tag{1.37}$$

Eq. 1.36 results from the orientation of molecular dipoles of species k with permanent dipole moment, $\vec{\mu}_k$, and number density, ρ_k, in the external field against their thermal motion. For species with molecular polarizability, α_k, Eq. 1.37 describes the induced polarization in the medium caused by the inner field, $(\hat{\vec{E}}_i)_k$, acting at the position of the molecule. As $\hat{\vec{P}}_\mu$ and $\hat{\vec{P}}_\alpha$ occur on different time scales, these two effects can be separated.[61] To summarize the magnitudes of induced polarization effects occurring at infrared to ultraviolet frequencies, an infinite frequency permittivity, ε_∞, is introduced via

$$\hat{\vec{P}}_\mu = \varepsilon_0 (\hat{\varepsilon} - \varepsilon_\infty) \hat{\vec{E}} \tag{1.38}$$

$$\hat{\vec{P}}_\alpha = \varepsilon_0 (\varepsilon_\infty - 1) \hat{\vec{E}} \tag{1.39}$$

where $\hat{\vec{P}}_\mu$ reflects all contributions that depend on frequency, irrespective of their rotational, (inter- and intramolecular) vibrational, librational (restricted rotations), or translational character.[38] Characteristic times of these processes are in the order of femto- to nanoseconds. Thus, measurement of $\hat{\varepsilon}(\omega)$ in the MHz to THz region provides insights into the dynamics of liquids.

1.2.2 Response functions of the orientational polarization

At low frequencies, the molecular dipoles are able to follow the variation of an oscillating electric field without delay. When the field frequency becomes sufficiently high, an instantaneous response is not possible any more and the polarization cannot reach its maximum. To describe the behavior of the orientational polarization for a time-dependent field, $\hat{\vec{P}}$ is related to $\hat{\vec{E}}$ by introduction of response functions.

For small enough $\hat{\vec{E}}$, a linear medium can be assumed, meaning that if a field $\vec{E}_1$ generates a polarization $\vec{P}_1$ and field $\vec{E}_2$ a polarization $\vec{P}_2$, then the field $\vec{E}_1 + \vec{E}_2$ results in a polarization $\vec{P}_1 + \vec{P}_2$. Consider an isotropic linear dielectric material that is polarized by an electric field. At time $t = 0$, the field is switched off and the time evolution of the polarization is recorded.

The induced polarization will follow changes of the applied field without delay, whereas the orientational polarization can be written as

$$\hat{\vec{P}}_\mu(t) = \hat{\vec{P}}_\mu(0) \cdot F_P^{or}(t) \tag{1.40}$$

where $F_P^{or}(t)$ is called the response or decay function of the polarization. It is defined as

$$F_P^{or}(t) = \frac{\langle \vec{P}_\mu(0) \cdot \vec{P}_\mu(t) \rangle}{\langle \vec{P}_\mu(0) \cdot \vec{P}_\mu(0) \rangle}. \tag{1.41}$$

For $t = 0$ it follows that $F_P^{or}(0) = 1$; for high values of t, $\hat{\vec{P}}$ will reach the equilibrium value and consequently $F_P^{or}(\infty) = 0$.
For a harmonic electric field, $\hat{\vec{E}}(t) = \hat{\vec{E}}_0 \exp(-i\omega t)$, the orientational polarization is defined as

$$\hat{\vec{P}}(\omega, t) = \varepsilon_0 (\varepsilon - \varepsilon_\infty) \hat{\vec{E}}(t) \mathcal{L}_{i\omega}[f_P^{or}(t')] \quad \text{with} \tag{1.42}$$

$$\mathcal{L}_{i\omega}[f_P^{or}(t')] = \int_0^\infty \exp(-i\omega t') f_P^{or}(t') dt'. \tag{1.43}$$

Here, $\mathcal{L}_{iw}[f_P^{or}(t')]$ is the Laplace-transformed pulse response function of the orientational polarization, which is connected to F_P^{or}, via

$$f_P^{or}(t') = -\frac{\partial F_P^{or}(t-t')}{\partial (t-t')} \quad \text{and normalized, i.e.} \quad \int_0^\infty f_P^{or}(t') dt' = 1. \tag{1.44}$$

The complex permittivity, $\hat{\varepsilon}(\omega)$, can than be calculated as[60]

$$\hat{\varepsilon}(\omega) = \varepsilon'(\omega) - i\varepsilon''(\omega) = \varepsilon_\infty + (\varepsilon - \varepsilon_\infty) \cdot \mathcal{L}_{i\omega}[f_P^{or}(t')]. \tag{1.45}$$

1.3 Empirical description of dielectric relaxation

To characterize the behavior of the orientational polarization, a number of equations are used to describe the experimental data.

1.3.1 Debye equation

The Debye (D) equation can be obtained by assuming that the decrease of the polarization in the absence of an electric field is directly proportional to the polarization itself.[62,63] Then, the polarization is described by

$$\frac{\partial}{\partial t} \vec{P}_\mu(t) = -\frac{1}{\tau} \vec{P}_\mu(t), \tag{1.46}$$

1.3. EMPIRICAL DESCRIPTION OF DIELECTRIC RELAXATION

where τ is the relaxation time. From the solution of Eq. 1.46,

$$\vec{P}_\mu(t) = \vec{P}_\mu(0) \exp\left(-\frac{t}{\tau}\right), \tag{1.47}$$

the pulse response function (Eq. 1.48) is obtained as

$$f_P^{\text{or}}(t) = \frac{1}{\tau} \exp\left(-\frac{t}{\tau}\right). \tag{1.48}$$

Applying Eq. 1.45, the Debye equation for the complex dielectric permittivity is obtained as

$$\hat{\varepsilon}(\omega) = \varepsilon_\infty + \frac{\varepsilon - \varepsilon_\infty}{1 + i\omega\tau}. \tag{1.49}$$

The dispersion and loss curves are

$$\varepsilon'(\omega) = \varepsilon_\infty + \frac{\varepsilon - \varepsilon_\infty}{1 + \omega^2\tau^2} \quad \text{and} \tag{1.50}$$

$$\varepsilon''(\omega) = \omega\tau \frac{\varepsilon - \varepsilon_\infty}{1 + \omega^2\tau^2}. \tag{1.51}$$

On a logarithmic scale, the real part is a monotonically decreasing point-symmetric function and the imaginary part is a symmetric peak with a maximum value at $\omega = 2\pi\nu = 1/\tau$.

1.3.2 Distribution functions

A distribution of relaxation times on a linear, $g(\tau)$, or logarithmic scale, $G(\ln \tau)$, may be used for the description of dielectric relaxation.[60] The complex permittivity is written as

$$\hat{\varepsilon}(\omega) = \varepsilon_\infty + (\varepsilon - \varepsilon_\infty) \int_0^\infty \frac{G(\ln \tau)}{(1 + i\omega\tau)} d\ln\tau \quad \text{with} \quad \int_0^\infty G(\ln\tau) d\ln\tau = 1. \tag{1.52}$$

As the distribution functions cannot be determined from the experimental spectra in a straightforward way,[60] empirical extensions of the Debye equation have been introduced.

Cole-Cole equation. A symmetrically-broadened loss curve in combination with a flatter dispersion curve is modelled by a Cole-Cole (CC) equation,[64,65]

$$\hat{\varepsilon}(\omega) = \varepsilon_\infty + \frac{\varepsilon - \varepsilon_\infty}{1 + (i\omega\tau)^{1-\alpha}}. \tag{1.53}$$

The CC parameter, $\alpha \in [0..1[$, describes a symmetric relaxation time distribution of the principal relaxation time, τ. For $\alpha = 0$, Eq. 1.53 reduces to the Debye equation.

Modified Cole-Cole equation. To account for the inertial rise of the dipole reorientation, a modified Cole-Cole (CC^m) equation,

$$\hat{\varepsilon}(\omega) = \varepsilon_\infty + \frac{\varepsilon - \varepsilon_\infty}{1 - (1 + \gamma_{\text{lib}}\tau)^{-(1-\alpha)}} \cdot \left(\frac{1}{1 + (i\omega\tau)^{1-\alpha}} - \frac{1}{1 + (i\omega\tau + \gamma_{\text{lib}}\tau)^{1-\alpha}} \right) \quad (1.54)$$

may be used.[66] Eq. 1.54 avoids unphysical contributions of the CC equation at high (THz to far-infrared) frequencies, where librational and/or vibrational modes contribute to the spectra. The inertial rise constant, γ_{lib}, is in the order of the resonance frequency of the librational/vibrational mode(s). Note that for $\alpha = 0$ the corresponding modified Debye equation (D^m) is obtained.

Cole-Davidson equation. An asymmetrical relaxation time distribution is described by the Cole-Davidson (CD) equation,[67,68]

$$\hat{\varepsilon}(\omega) = \varepsilon_\infty + \frac{\varepsilon - \varepsilon_\infty}{(1 + i\omega\tau)^\beta}, \quad (1.55)$$

with the empirical CD parameter, $\beta \in]0..1]$. The Debye equation is obtained for $\beta = 1$.

Havriliak-Negami equation. For the representation of broadened and asymmetrically shaped dispersion and loss curves both parameters $\alpha \in [0..1[$ and $\beta \in]0..1]$ are combined in the Havriliak-Negami (HN) equation,[69]

$$\hat{\varepsilon}(\omega) = \varepsilon_\infty + \frac{\varepsilon - \varepsilon_\infty}{[1 + (i\omega\tau)^{1-\alpha}]^\beta} \quad (1.56)$$

For $\alpha = 0$ and $\beta = 1$, Eq. 1.56 is equal to the Debye equation.

1.3.3 Damped harmonic oscillator

Resonant absorptions, like vibrations and librations in the THz or far-infrared regions, can be modelled as a damped harmonic oscillator (DHO). Assuming a harmonic oscillator subjected to a damping force and driven by a harmonically oscillating field, one obtains

$$\hat{\varepsilon}(\omega) = \varepsilon_\infty + \frac{(\varepsilon - \varepsilon_\infty)\omega_0^2}{(\omega_0^2 - \omega^2) + i\omega\tau_D^{-1}}. \quad (1.57)$$

as the solution of the differential equation describing the time-dependent motion, $x(t)$, of an effective charge, q.[70,71] In Eq. 1.57, $\omega_0 = \sqrt{k/m} = 2\pi\nu_0$ and $\gamma = 1/(2\pi\tau_D)$ are the angular resonance frequency and damping constant of the oscillator, respectively. For $\tau_D \ll \omega_0^{-1}$, Eq. 1.57 reduces to the Debye equation.

1.3.4 Combination of models

In real systems, the DR spectrum may be the result of a superposition of distinct relaxation modes. Therefore, Eq. 1.52 is written as a sum of $j = 1 \ldots n$ separate processes:

$$\hat{\varepsilon}(\omega) = \varepsilon_\infty + \sum_{j=1}^{n} (\varepsilon_j - \varepsilon_{\infty,j}) \int_0^\infty \frac{G_j(\ln \tau_j)}{1 + i\omega\tau_j} \mathrm{d}\ln\tau_j \qquad (1.58)$$

Each process is characterized by its own relaxation time, τ_j, and dispersion amplitude, S_j, defined via:

$$\varepsilon - \varepsilon_\infty = \sum_{j=1}^{n} (\varepsilon_j - \varepsilon_{\infty,j}) = \sum_{j=1}^{n} S_j \qquad (1.59)$$

$$\varepsilon_{\infty,j} = \varepsilon_{j+1} \qquad (1.60)$$

This leads to the general expression for superpositions of HN, CC$^\mathrm{m}$ and DHO equations:

$$\begin{aligned}
\hat{\varepsilon}(\omega) = \varepsilon_\infty \; &+ \; \sum_j \frac{S_j}{[1 + (\mathrm{i}\omega\tau_j)^{1-\alpha_j}]^{\beta_j}} \\
&+ \; \sum_k \frac{S_k}{1 - (1 + \gamma_{\mathrm{lib},k}\tau_k)^{-(1-\alpha_k)}} \cdot \left(\frac{1}{1 + (\mathrm{i}\omega\tau_k)^{1-\alpha_k}} - \frac{1}{1 + (\mathrm{i}\omega\tau_k + \gamma_{\mathrm{lib},k}\tau_k)^{1-\alpha_k}} \right) \\
&+ \; \sum_l \frac{S_l \omega_{0,l}^2}{(\omega_{0,l}^2 - \omega^2) + \mathrm{i}\omega\tau_{\mathrm{D},l}^{-1}}
\end{aligned} \qquad (1.61)$$

1.3.5 Data processing

To extract physical information from complex permittivity spectra, an appropriate mathematical description of the measured complex permittivity data has to be found. As mentioned above, the dielectric response may be created by more than one relaxation process. In the ideal case, each process can be unambiguously modelled by one empirical function. However, due to the broad nature of the relaxations and technological limitations,[38] a decomposition is rarely trivial. Therefore, more than one relaxation model can possibly describe the experimental spectra and thus, the choice of the 'true' relaxation model has to follow some rules. First of all, the parameters obtained have to be physically meaningful. Second, the normalized variance of the fit, χ_r^2, defined as

$$\chi_\mathrm{r}^2 = \frac{1}{2N - m - 1} \left[\sum_{i=1}^{N} w_{\varepsilon'}(\nu_i)\delta\varepsilon'(\nu_i)^2 + \sum_{i=1}^{N} w_{\varepsilon''}(\nu_i)\delta\varepsilon''(\nu_i)^2 \right], \qquad (1.62)$$

should be small. In Eq. 1.62, $\delta\varepsilon'(\nu_i)$ and $\delta\varepsilon''(\nu_i)$ are the residuals, N is the number of data triples $[\nu_i, \varepsilon'(\nu_i), \varepsilon''(\nu_i)]$ and m the number of the adjustable parameters; $w_{\varepsilon'}(\nu_i)$ and $w_{\varepsilon''}(\nu_i)$

are the weights; only unweighted fits, $w_{\varepsilon'}(\nu_i) = w_{\varepsilon''}(\nu_i) = 1$, were performed in the analysis of the spectra presented here. Figure 1.1 shows the effect of a superposition of three (3D model) and four (4D model) Debye equations to describe a spectrum that was simulated by a combination of a CC and a D equation (CC + D model). Apart from model-dependent fluctuations, the relative percentage deviations, $\delta\varepsilon'_{\text{fit}} = 100 \cdot (\varepsilon'_{\text{CC+D}} - \varepsilon'_{\text{3D,4D}})/\varepsilon'_{\text{CC+D}}$ and $\delta\varepsilon''_{\text{fit}} = 100 \cdot (\varepsilon''_{\text{CC+D}} - \varepsilon''_{\text{3D,4D}})/\varepsilon''_{\text{CC+D}}$, for the real and imaginary parts are well below the probable experimental uncertainties of ca. $\pm 2\,\%$ in $\hat{\varepsilon}$.[38] Thus, consideration of χ^2_r alone will not necessarily provide meaningful physical insights.

Additionally, the number of parameters should be as small as possible and the relaxation model should not change in a concentration series, except for specific physical reasons.

The overall fitting procedure performed in the present PhD thesis was the following: experimental $\hat{\eta}(\nu)$ data were corrected for the conductivity contribution. Then, different relaxation models were tested by simultaneously fitting $\varepsilon'(\nu)$ and $\varepsilon''(\nu)$ using the MWFIT program. A nonlinear least-squares routine based on the method of Levenberg and Marquardt is implemented in this program.[72] To minimize systematic deviations at low ν, the conductivity was slightly varied. The origin of the small deviations of the resulting corrected values from conventionally measured (dc) conductivities are well understood.[37]

Due to the nonlinear nature of the fitting process, it is not possible to assign statistically meaningful standard uncertainties to the individual fit parameters, but the square root of the diagonal elements of the covariance matrix can be used as a measure for the certainty of the resulting parameters.[72,73]

1.4 Microscopic models of dielectric relaxation

1.4.1 Onsager equation

To describe the dielectric relaxation of liquid systems, Onsager[60,74] assumed spherical particles, which are embedded in a dielectric continuum and do not show any specific molecular interactions. Onsager deduced the equation

$$\varepsilon_0(\varepsilon - 1)\vec{E} = \vec{E}_h \cdot \sum_j \frac{\rho_j}{1 - \alpha_j f_j} \left(\alpha_j + \frac{1}{3k_\text{B}T} \cdot \frac{\mu_j^2}{1 - \alpha_j f_j} \right) \quad (1.63)$$

to connect macroscopic (ε) and microscopic (the polarizability, α_j, and the dipole moment, μ_j, of molecular-level species j) properties. In Eq. 1.63, ρ_j represents the charge density and f_j the reaction field factor describing a spherical cavity of finite radius, in which the particle is embedded. Note, that the Onsager equation is only valid for systems with a single dispersion step.

1.4. MICROSCOPIC MODELS OF DIELECTRIC RELAXATION

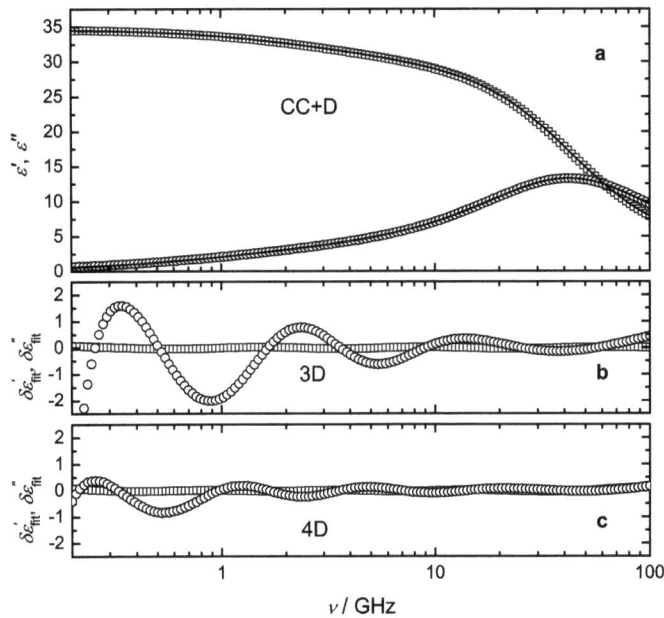

Figure 1.1: **(a)** Calculated dielectric permittivity, $\varepsilon'(\nu)$ (squares), and loss, $\varepsilon''(\nu)$ (circles), spectra for a CC + D model ($\varepsilon = 34.8$, $\tau_1 = 47.3$ ps, $\alpha = 0.23$, $\varepsilon_2 = 28.9$, $\tau_2 = 3.65$ ps, $\varepsilon_\infty = 3.75$), together with fits obtained using the 3D and 4D models (two full lines, which are essentially the same). **(b)** Relative deviations of calculated values from dispersion, $\delta\varepsilon'_{\text{fit}}$ (squares), and loss curves, $\delta\varepsilon''_{\text{fit}}$ (circles), obtained by fitting a 3D model to the calculated spectrum ($\varepsilon = 34.5$, $\tau_1 = 131$ ps, $\varepsilon_2 = 32.6$, $\tau_2 = 31.6$ ps, $\varepsilon_3 = 29.6$, $\tau_3 = 3.69$ ps, $\varepsilon_\infty = 3.88$, $\chi_r^2 = 0.00055$). **(c)** Corresponding values obtained by a 4D fit ($\varepsilon = 34.6$, $\tau_1 = 232$ ps, $\varepsilon_2 = 33.7$, $\tau_2 = 60.4$ ps, $\varepsilon_3 = 31.1$, $\tau_3 = 18.9$ ps, $\varepsilon_3 = 29.4$, $\tau_3 = 3.65$ ps, $\varepsilon_\infty = 3.84$, $\chi_r^2 = 0.000041$).

For a spherical cavity in a dielectric material, the cavity field, $\vec{E}_h$, is given by[60]

$$\vec{E}_h = \frac{3\varepsilon}{2\varepsilon + 1}\vec{E}, \qquad (1.64)$$

and thus the general form of the Onsager equation is obtained as

$$\frac{(\varepsilon - 1)(2\varepsilon + 1)\varepsilon_0}{3\varepsilon} = \sum_j \frac{\rho_j}{1 - \alpha_j f_j}\left(\alpha_j + \frac{1}{3k_\mathrm{B}T} \cdot \frac{\mu_j^2}{1 - \alpha_j f_j}\right). \qquad (1.65)$$

For a non-polarizable ($\alpha_j = 0$), dipolar liquid, this equation leads to the simplified form of the Onsager equation,

$$\frac{(\varepsilon - \varepsilon_\infty)(2\varepsilon + \varepsilon_\infty)}{\varepsilon(\varepsilon_\infty + 2)^2} = \frac{\rho\mu^2}{9\varepsilon_0 k_\mathrm{B}T}. \qquad (1.66)$$

To account for specific intermolecular interactions, statistical mechanics was applied by Kirkwood and Fröhlich.[75,76] Based on their theory, which included correlations among the dipoles, they derived the equation

$$\frac{(\varepsilon - \varepsilon_\infty)(2\varepsilon + \varepsilon_\infty)}{\varepsilon(\varepsilon_\infty + 2)^2} = \frac{\rho\mu^2}{9\varepsilon_0 k_\mathrm{B}T} \cdot g_\mathrm{K}, \qquad (1.67)$$

where g_K is the so-called Kirkwood factor. It is a measure of the interactions among the dipoles. That is, it represents orientational correlations, with $g_\mathrm{K} > 1$ corresponding to preferentially parallel orientations and $g_\mathrm{K} < 1$ to antiparallel orientations. The value $g_\mathrm{K} = 1$ implies a statistical arrangement of dipoles.

1.4.2 Cavell equation

Going one step further, Cavell[77] extended the Onsager equation (Eq. 1.65) to systems with more than a single dispersion step. The Cavell equation

$$\frac{\varepsilon + A_j(1 - \varepsilon)}{\varepsilon} \cdot S_j = \frac{N_\mathrm{A} c_j}{3k_\mathrm{B}T\varepsilon_0} \cdot \mu_{\mathrm{eff},j}^2 \qquad (1.68)$$

connects the dispersion amplitude, $S_j = \varepsilon_j - \varepsilon_{j+1}$, of relaxation process j to the molar concentration of the species, c_j, and their effective dipole moments, $\mu_{\mathrm{eff},j}$. The shape factor A_j accounts for the shape of the relaxing particle; for spheres, $A_j = 1/3$, but it can be calculated for ellipsoids of any shape (half-axes $a_j > b_j > c_j$) via the equation[60,78]

$$A_j = \frac{a_j b_j c_j}{2}\int_0^\infty \frac{\mathrm{d}s}{(s + a_j^2)^{3/2}(s + b_j^2)^{1/2}(s + c_j^2)^{1/2}}. \qquad (1.69)$$

An expression for prolate ellipsoids ($b_j = c_j$) was derived by Scholte,[79]

$$A_j = -\frac{1}{p_j^2 - 1} + \frac{p_j}{(p_j^2 - 1)^{1.5}}\ln\left(p_j + \sqrt{p_j^2 - 1}\right) \qquad \text{with} \qquad p_j = \frac{a_j}{b_j} \qquad (1.70)$$

The values of $\mu_{\text{eff},j}$ are connected to the apparent dipole moment of the species, $\mu_{\text{ap},j}$, i.e. the dipole moment in absence of orientational correlations, via

$$\mu_{\text{eff},j} = \sqrt{g_j}\mu_{\text{ap},j} \tag{1.71}$$

Here, the (empirical) factor g_j is a measure for the strength of the correlations whose values are interpreted as for the Kirkwood factor g_K (Eq. 1.67). Inclusion of cavity- and reaction-field effects yields

$$\mu_{\text{ap},j} = \frac{\mu_j}{1 - f_j \alpha_j} \tag{1.72}$$

as an expression for connecting the dipole moment of an isolated gas phase species (μ_j) to an uncorrelated state in solution ($\mu_{\text{ap},j}$). The reaction field factor f_j can be calculated for a spherical cavity of radius a_j via[60]

$$f_j = \frac{1}{4\pi\varepsilon_0 a_j^3} \cdot \frac{2\varepsilon - 2}{2\varepsilon + 1} \tag{1.73}$$

or, more generally, for ellipsoidal particles via[80]

$$f_j = \frac{3}{4\pi\varepsilon_0 a_j b_j c_j} \cdot \frac{A_j(1 - A_j)(\varepsilon - 1)}{\varepsilon + (1 - \varepsilon)A_j}. \tag{1.74}$$

1.4.3 Debye model of rotational diffusion

According to the Debye model of rotational diffusion, a system consists of particles rotating freely in space. Collisions among the particles are frequent and thus cause a reorientation of the dipoles (the so-called diffusion of dipole orientation).[62]

However, Debye's theory is based on a number of assumptions: for the reorientation of spherical particles, inertial effects and dipole-dipole interactions are neglected, and it is assumed that the hydrodynamic laws of rotation of macroscopic particles in a liquid can be applied on the microscopic level.[62] As a consequence, the theory is only valid for non-associating systems and particles that are large compared to their surrounding ones.[81]

By using the Lorentz field as the inner field, Debye obtained the dipole correlation function,[60]

$$\gamma(t) = \exp\left(-\frac{t}{\tau'}\right). \tag{1.75}$$

The microscopic relaxation time, τ' (see below), is related to the friction factor, ζ, and to the microscopic viscosity, η', i.e. the dynamic viscosity of the environment of the sphere, via the Stokes-Einstein-Debye (SED) equation:

$$\tau' = \frac{\zeta}{2k_B T} = \frac{3V_m \eta'}{k_B T}. \tag{1.76}$$

Here, V_m is the molecular volume of the rotating sphere.

However, the connection between microscopic and macroscopic, η, viscosities is not clear. To circumvent this problem, some additional parameters are introduced in Eq. 1.76. A frequently used expression is given by Dote et al.[82]

$$\tau' = \frac{3V_\mathrm{m}\eta}{k_\mathrm{B}T} fC + \tau'^{,0} \tag{1.77}$$

The experimentally found axis-intercept is treated by the empirical parameter $\tau'^{,0}$, which is sometimes associated with the free-rotator correlation time. The shape factor, f, is a purely geometrical parameter, which accounts for the deviation of the shape of the molecule from that of a sphere. For prolate bodies with axial symmetry it was found that[83,84]

$$f = \frac{\frac{2}{3}[1-(\alpha^\perp)^4]}{\frac{[2-(\alpha^\perp)^2](\alpha^\perp)^2}{[1-(\alpha^\perp)^2]^{1/2}} \ln\left[\frac{1+[1-(\alpha^\perp)^2]^{1/2}}{\alpha^\perp}\right] - (\alpha^\perp)^2}, \tag{1.78}$$

where $\alpha^\perp$ is the ratio of the particle volume and the volume swept out as the particle rotates about an axis perpendicular to the symmetry axis through the center of hydrodynamic stress.[84] For a prolate ellipsoid with major half-axis a and minor half-axis b, $\alpha^\perp = b/a$ may be assumed.[84] The friction parameter, C, represents a correction of the difference between macroscopic and microscopic viscosities. Its limiting values are $C = 1$ for stick and $C = 1 - f^{-2/3}$ for slip boundary conditions. However, under special conditions, for example the rotation of very small molecules, values of $C < C_\mathrm{slip}$ are possible.[85] In cases where values of f and C cannot be determined independently, the discussion is sometimes limited to effective volumes of rotation, defined as

$$V_\mathrm{eff} = fCV_\mathrm{m}. \tag{1.79}$$

1.4.4 Microscopic and macroscopic relaxation times

DR measurements probe the collective dynamics of a system and therefore the macroscopic relaxation time, τ, is determined. To allow comparison with other techniques and to interpret DR spectra on a molecular level, it is necessary to connect macroscopic and microscopic, τ', relaxation times. A number of theoretical approaches exists, but the most commonly used is given by Powles and Glarum,[86,87]

$$\tau = \frac{3\varepsilon}{2\varepsilon + \varepsilon_\infty} \cdot \tau' \tag{1.80}$$

This equation is valid for pure rotational diffusion. A more generalized form is given by Madden and Kivelson,[88]

$$\tau = \frac{3\varepsilon}{2\varepsilon + \varepsilon_\infty} \cdot \frac{g_\mathrm{K}}{\dot{g}} \cdot \tau' \tag{1.81}$$

1.5. ION PAIR RELAXATION

where the Kirkwood factor, g_K, and the dynamic correlation factor, $\dot{g}$, account for dipole-dipole correlations. For the limit $g_K/\dot{g} = 1$ Eq. 1.81 reduces to the Powles-Glarum equation (Eq. 1.80).

1.5 Ion pair relaxation

All electrolyte solutions show some tendency to form ion pairs, i.e. to associate. The extent of ion pairing increases with ionic charge and with decreasing solvent permittivity. DR spectroscopy is particularly sensitive to the presence of ion pairs in solution.[34] Among the possible species, contact ion pairs (CIPs), solvent-shared ion pairs (SIPs) or double solvent separated ion pairs (2SIPs), can contribute to DR spectra.[43]

1.5.1 Amplitudes

For systems showing a distinct ion pair dispersion, Eqs. 1.68-1.74 (with $S_j = S_{IP}$) are used to determine the concentration of ion pairs, c_{IP}, in solution. For the calculation of A_{IP} and μ_{IP}, geometrical parameters, the polarizability and the gas phase dipole moment of the corresponding species have to be available.

The situation is more complicated, when the ion pair relaxation is overlapping with a relaxation located close to it. Consider the equilibrium between free cations, C^+, and anions, A^-:

$$C^+ + A^- \underset{k_{-1}}{\overset{k_1}{\rightleftharpoons}} [IP]^0 \qquad (1.82)$$

with rate constants of ion pair (IP) formation, k_1, and decay, k_{-1}, where

$$K_A^\circ = k_1/k_{-1} \qquad (1.83)$$

is the standard (infinite dilution) association constant of the ion pair.

Assuming two contributing species, here C^+ and IP, the additivity of their amplitudes and a spherical shape of the particles, Eq. 1.68 expresses the experimentally observed amplitude for that relaxation, S, by the equation:

$$S = \frac{\varepsilon}{2\varepsilon + 1} \cdot \frac{N_A}{3k_B T \varepsilon_0} \cdot \left(c_{C^+} g_{C^+} \mu_{ap,C^+}^2 + c_{IP} g_{IP} \mu_{ap,IP}^2 \right) \qquad (1.84)$$

For small ion pair concentrations, $g_{IP} = 1$ is fulfilled.[44] When the values of μ_{ap,C^+}, g_{C^+} and $\mu_{ap,IP}$ are known (see below) and the molar concentrations of cations, c_{C^+}, and ion pairs, c_{IP}, are connected by

$$c = c_{C^+} + c_{IP}, \qquad (1.85)$$

where c is the analytical molar concentration of the salt formed by C^+ and A^-, Eq. 1.84 yields the concentrations of the contributing species together with the association constants,

$$K_A = c_{IP}/(c - c_{IP})^2. \tag{1.86}$$

These can be used to estimate K_A° by extrapolation with a Guggenheim-type equation:[44]

$$\log K_A = \log K_A^\circ - \frac{2A_{DH}\sqrt{I}}{1 + R_{ij}B_{DH}\sqrt{I}} + A_K I + B_K I^{3/2} \tag{1.87}$$

where I ($\equiv c$ for 1:1 electrolytes) is the stoichiometric ionic strength, A_{DH} and B_{DH} are the Debye-Hückel coefficients and R_{ij} is the upper limit of the distance at which the ions are considered to be associated, which can be calculated from the ionic radii of the anions, cations and the length of an orientated solvent molecule;[54,89] Y_K ($Y = A, B$) are adjustable parameters.

1.5.2 Kinetics

Formation and decay of ion pairs can be described by the chemical equilibrium given in Eq. 1.82. Small perturbations cause a fluctuation of the ion pair concentration, x, defined via

$$c_{C^+} = c_{C^+}^0 + x \tag{1.88}$$

$$c_{A^-} = c_{A^-}^0 + x \tag{1.89}$$

$$c_{IP} = c_{IP}^0 - x \tag{1.90}$$

where c_j^0 are the corresponding equilibrium concentrations. For small x, the rate equation

$$\frac{dx}{dt} = -k_1(c_{C^+}^0 + x)(c_{A^-}^0 + x) + k_{-1}(c_{IP}^0 - x) \approx -\left(k_1(c_{C^+}^0 + c_{A^-}^0) + k_{-1}\right)x \tag{1.91}$$

is valid. With the common formalism of relaxation kinetics[90] one obtains as the solution of Eq. 1.91

$$x = x_0 \exp\left(-\frac{t}{\tau_{ch}}\right) \quad \text{with} \tag{1.92}$$

$$\tau_{ch}^{-1} = k_1(c_{C^+}^0 + c_{A^-}^0) + k_{-1} \tag{1.93}$$

where $x_0 = x(t = 0)$. Here, τ_{ch}^{-1} is the chemical relaxation rate and τ_{ch} is the system's equilibration time.

The observable relaxation rate, τ_{IP}^{-1}, is given by the sum of chemical and orientational relaxation rates,

$$\frac{1}{\tau_{IP}} = \frac{1}{\tau'_{IP}} + \frac{1}{\tau_{ch}} = \frac{1}{\tau'_{IP}} + k_{-1} + k_1(c_{C^+}^0 + c_{A^-}^0) \tag{1.94}$$

$$= \frac{1}{\tau'_{IP}} + k_{-1} + 2k_1(c - c_{IP}) \tag{1.95}$$

1.6. TEMPERATURE DEPENDENCE OF RELAXATION TIMES

Eqs. 1.83–1.95 can be used to determine k_1, k_{-1} and K_A° from a linear fit of $\tau_{\text{IP}}^{-1} = f(c - c_{\text{IP}})$. From the value of τ_{IP}', the effective volumes of rotation of the ion pair can be estimated using Eqs. 1.77 & 1.79.

Assuming diffusion-controlled processes in which Coulomb interactions are dominant, the rate constants of ion pair formation, k_1^D, and decay, k_{-1}^D, follow Eigen's theory:[91]

$$k_1^D = \frac{N_A z_{C^+} z_{A^-} e_0^2}{\varepsilon_0 \varepsilon k_B T} \cdot \frac{D_{C^+} + D_{A^-}}{\exp\left[\frac{z_{C^+} z_{A^-} e_0^2}{4\pi\varepsilon_0 \varepsilon k_B T d}\right] - 1} \tag{1.96}$$

$$k_{-1}^D = \frac{3 z_+ z_- e_0^2}{4\pi\varepsilon_0 \varepsilon k_B T d^3} \cdot \frac{D_+ + D_-}{1 - \exp\left[\frac{-z_+ z_- e_0^2}{4\pi\varepsilon_0 \varepsilon k_B T d}\right]} \tag{1.97}$$

where d is the distance between cation and anion in the ion pair and z_i ($i = C^+, A^-$) are their respective charges. The diffusion coefficients of the ions, D_i, can be estimated from the single ion conductivities, λ_i^∞:[92]

$$D_i = \frac{RT}{|z_i| F^2} \lambda_i^\infty, \tag{1.98}$$

where F is Faraday's constant and R is the universal gas constant.

1.6 Temperature dependence of relaxation times

Arrhenius equation. The Arrhenius equation was originally introduced to describe the variation of the rate constant of a chemical reaction with temperature. It is based on the following model: in order to transform reactants into products, they first need to acquire a minimum amount of energy, called the activation energy E_a. The (empirical) application of this model to relaxation times has the form[93]

$$\ln \tau = \ln \tau_0 + \frac{E_a}{RT} \tag{1.99}$$

The preexponential factor, τ_0, called the frequency factor, is interpreted as the shortest possible relaxation time.

Eyring equation. Eyring applied transition state theory to express the temperature dependence of rate processes.[94] He introduced the Gibbs energy of activation,

$$\Delta G^{\neq} = \Delta H^{\neq} - T\Delta S^{\neq}, \tag{1.100}$$

with the corresponding enthalpy, $\Delta H^{\neq}$ and entropy, $\Delta S^{\neq}$, components. Finally he obtained

$$\ln \tau = \ln \frac{h}{k_B T} - \frac{\Delta S^{\neq}}{R} + \frac{\Delta H^{\neq}}{RT} \tag{1.101}$$

where h is Planck's constant. This equation neglects the possible temperature dependencies of $\Delta H^{\neq}$ and $\Delta S^{\neq}$. Nevertheless, it was often found to be valid for small temperature ranges, where the temperature dependence of $\ln(h/k_B T)$ can be neglected. Both the Arrhenius and Eyring equations are equally suitable to describe the linearity of plots of $\ln \tau = f(1/T)$. The parameters of Eqs. 1.99 & 1.101 are connected via

$$E_a = \Delta H^{\neq} - RT \ln T \quad \text{and} \tag{1.102}$$

$$\ln \tau_0 = \ln \frac{h}{k_B} - \frac{\Delta S^{\neq}}{R} \tag{1.103}$$

Eq. 1.101 was used to fit the values of $\tau_1 = f(T)$ of neat acetonitrile (Section 3.2).

Vogel-Fulcher-Tammann equation. Glass-forming liquids are often characterized by a nonlinearity in the $\ln \tau = f(1/T)$ representation. The theory is based on the free volume, which is required to allow reorientation of a molecule. The free volume is defined as the difference between the macroscopic volume and the thermal volume of a particle.[95] Angell presented the Vogel-Fulcher-Tammann (VFT) equation in the form[96]

$$\ln(Y) = \ln(A_Y) + \frac{B_Y}{T - T_{0,Y}}, \tag{1.104}$$

which describes transport properties, $Y (= \kappa, \eta^{-1}, \ldots)$, of glass-forming liquids above their glass-transition temperature, T_g, over a wide temperature range. In Eq. 1.104, A_Y and B_Y are fit parameters and $T_{0,Y}$ is the so-called VFT-temperature, which is often found to be $\sim 20-30$ K below T_g determined via differential scanning calorimetry.[96] Generally, $T_{0,Y}$ is equal to the Kautzmann temperature, which is defined by the intersection of the entropy curve of the liquid and the solid. The ratio $B_Y/T_{0,Y}$ is defined as the so-called strength parameter.[96]

Eq. 1.104 was applied in fitting the conductivities ($Y = \kappa$) of neat ionic liquids over a wide temperature range, which have been recently published in ref. 97.

Chapter 2

Experimental

2.1 Materials

2.1.1 Solvents

High-purity solvents were used throughout. The following list contains further purification and purity details.

Acetonitrile (AN, Merck $\geq$ 99.9 %) was distilled over CaH_2 (Merck > 95 %) in a purpose-built circulation apparatus and subsequently stored over activated 4 Å molecular sieves. The GC purity (Hewlett Packard 6890 Series GC system) of the product was > 99.99 %, and coulometric Karl Fischer titration (Mitsubishi Moisturemeter MCI CA-02) yielded a H_2O mass fraction < 50 ppm.

Dichloromethane (DCM, Acros > 99.9 %), dimethylsulfoxide (DMSO, Merck > 99.5 %) and propylene carbonate (PC, Sigma-Aldrich 99.7 %) were stored over activated 4 Å molecular sieves. These solvents had GC purities > 99.94 % for PC and > 99.99 % for DMSO and DCM. Their water mass fractions were < 20 ppm.

Benzonitrile (BN, Sigma-Aldrich > 99.9 %), 1-butanol (1-BuOH, Riedel-de Haën > 99.5 %) and N,N-dimethylacetamide (DMA, Fluka > 99.8 %) were stored over activated 4 Å molecular sieves, but otherwise used as received.

Methanol (MeOH, Merck $\geq$ 99.9 %) was distilled over Mg/I_2 (Merck > 99 % and > 99.5 %, respectively; ~6 % (w/w) I_2)[98] yielding > 99.99 % GC purity and mass fraction H_2O < 20 ppm.

Water was purified with a Millipore MILLI-Q purification unit, yielding batches with specific resistivity $\geq$ 18 MΩ cm^{-1}.

2.1.2 Ionic liquids

Although the situation has changed in recent years, the ionic liquids offered by suppliers some years ago were either expensive or of questionable (i.e. unstated) quality. The synthesis of most ILs (see below) is straightforward. It was therefore decided to synthesize the target ILs in our laboratory, whenever it was economically efficient (costs of commercial ILs vs. man-power) to do so. Highly pure products can only be obtained when starting materials are thoroughly purified and an atmosphere of inert gas is maintained during reaction.

Materials. The chemicals used for preparing the ILs were the following: 1-methylimidazole (mim, Merck & Carl Roth 99 %) was distilled over KOH under reduced pressure, stored over activated 4 Å molecular sieves and redistilled under reduced pressure prior to use. Alkyl halides (RHal)—1-bromoethane (Merck $\geq$ 99 %), 1-chlorobutane (Merck $\geq$ 99 %), 1-bromobutane (Merck $\geq$ 98 %), 1-chlorohexane (Merck $\geq$ 99 %) and 1-chlorooctane (Aldrich $\geq$ 99 %)—were distilled prior to use.
The salts $AgBF_4$ (Fluorochem, 99 %), $NaBF_4$ (VWR Prolabo 98.6 %), and KPF_6 (Fluorochem, 99 %) were dried but otherwise used as received, whereas sodium dicyanamide (NaDCA, Fluka $\geq$ 96 %, yellowish) was recrystallized from methanol yielding a colorless product.
The ILs [emim][BF_4] (> 98 %), [emim][$EtSO_4$] (99 %), [emim][DCA] (> 98 %) and [hmim][NTf_2] (99 %) were purchased from IoLiTec. These slightly yellowish ILs were dried and stored like the other ILs (see below) but otherwise used as received. The commercial [emim][BF_4] sample is further on designated as [emim][BF_4]#1.

Synthesis. In the first step, alkylmethylimidazolium halides ([Rmim][Hal]) were formed in an S_N2 reaction[99] from mim and the appropriate RHal. The target ILs were prepared from [Rmim][Hal] and the appropriate metal salt via anion metathesis. Except for [emim][BF_4] and [bmim][DCA], previously published routes were followed.[100–103]

The ILs [emim][Br], [bmim][Cl], [bmim][Br], [hmim][Cl] and [omim][Cl] were obtained by adding a slight molar excess, $n_{RHal} \approx 1.1 n_{mim}$, of the alkyl halide to a stirred solution ($\sim$40 % v/v) of mim in AN under an atmosphere of dry nitrogen. Complete conversion was checked by ^{1}H-NMR analysis. The alkylmethylimidazolium halides [emim][Br], [bmim][Cl] and [bmim][Br] so formed were recrystallized at least four times from AN to give colorless products, whereas [hmim][Cl] and [omim][Cl] were washed twice with ethyl acetate (Merck p.a.) yielding yellowish products. The solvent was removed at a vacuum line and the alkylmethylimidazolium halides were dried in vacuo. No impurities were detected with ^{1}H- and ^{13}C-NMR analyses.

The IL [emim][BF_4] was obtained via anion metathesis from equimolar amounts of the salts

2.1. MATERIALS

[emim][Br] and NaBF$_4$ dissolved in acetonitrile. After evaporation of the solvent on a vacuum line, dichloromethane was added in excess. The precipitated NaBr was separated by filtration and DCM subsequently removed by distillation. This procedure yielded [emim][BF$_4$] with 1.6 % (w/w) Br$^-$ impurity determined by potentiometric titration of an aqueous solution of the IL against a standard aqueous solution of AgNO$_3$ (Carl Roth). By adding equimolar amounts of AgBF$_4$ dissolved in MeOH the excess Br$^-$ was precipitated as AgBr and removed by filtration. Then the mixture was kept overnight at ca. $-18\,°C$, leading to phase separation into a yellowish methanol-rich phase and an IL-rich phase, which was isolated. This newly developed procedure is cheap and easy to perform, yields a pure and colorless product and circumvents the known difficulties in synthesizing highly-pure [emim][BF$_4$].[22] This sample is further on designated as [emim][BF$_4$]#2.

Equimolar amounts of the appropriate alkylmethylimidazolium halide dissolved in DCM were reacted with aqueous solutions of NaBF$_4$ and KPF$_6$ to produce [bmim][BF$_4$], [bmim][PF$_6$], [hmim][BF$_4$] and [omim][BF$_4$]. The aqueous phase was extracted three times with DCM. The organic fractions were combined and subsequently washed thrice with small amounts of water to remove traces of metal halides (NaCl, NaBr, KCl) before distilling off the DCM. The overall procedure yielded colorless ([bmim][BF$_4$] and [bmim][PF$_6$]) or yellowish ([hmim][BF$_4$] and [omim][BF$_4$]) products. All preparation steps involving aqueous phases were performed rapidly with materials cooled to $\sim 0\,°C$ prior to use to minimize the hydrolysis of BF$_4^-$ and PF$_6^-$.[22]

The IL [bmim][BF$_4$] was obtained either from the bromide or chloride salt. These two samples are further on designated as [bmim][BF$_4$]#1 and [bmim][BF$_4$]#2, respectively.

Equimolar amounts of [bmim][Br] and NaDCA were stirred at $\sim 35\,°C$ over night to give [bmim][DCA]. To separate the ionic liquid, an excess of DCM was added and the precipitated NaCl filtered off. After removal of the solvent under vacuum this procedure was repeated, yielding a slightly yellowish product.

Prior to use all ILs were dried under high-vacuum ($p < 10^{-8}$ bar) for 7 days at $\sim 40\,°C$. Water contents were determined by coulometric Karl Fischer titration and halide content was quantified by potentiometric titration. The levels of these two impurities (H$_2$O and Hal$^-$) are listed in Table 2.1. Additionally, ^{1}H-, ^{13}C- and ^{19}F-NMR spectra (where applicable) were recorded. Except for [emim][BF$_4$]#1, where a signal arising from an acidic proton (~ 0.01 mole fraction of impurity) was observed with a chemical shift of ~ 6.5 ppm, no contaminants were detected.

Table 2.1: Water, w_{H_2O}, and Halide Mass Fractions, w_{Hal}, Detected by Coulometric Karl Fischer and Potentiometric Precipitation Titrations in Samples of the Investigated Ionic Liquids.

IL	$w_{H_2O}/10^{-6}$	$w_{Hal}/10^{-6}$
[emim][BF$_4$]#1	50	bdl[a]
[emim][BF$_4$]#2	50	bdl
[emim][EtSO$_4$]	20	bdl
[emim][DCA]	40	400
[bmim][Cl]	800[b]	-
[bmim][BF$_4$]#1	40	150
[bmim][BF$_4$]#2	50	20
[bmim][PF$_6$]	50	20
[bmim][DCA]	100	5000
[hmim][BF$_4$]	50	20
[hmim][NTf$_2$]	20	100[c]
[omim][BF$_4$]	50	50

[a] below detection limit; [b] estimated by measuring a solution of [bmim][Cl] in AN ($w_{[bmim][Cl]} = 0.179$); [c] according to certificate of analysis.

2.1.3 Sample preparation

All dried ILs were stored in an N_2-filled glovebox. The N_2-protection was also maintained when preparing the investigated mixtures (either by the use of septum-sealed glass vials or by preparation of the mixtures in a purpose-built glovebox equipped with a device for subsequent solvent removal) and during all steps of sample handling, including the measurements. Mixtures were prepared immediately before use on an analytical balance without buoyancy correction and thus have a standard uncertainty of about $\pm 0.2\,\%$. In between successive measurements samples were stored in a desiccator to minimize possible uptake of ambient moisture.

2.2 Measurement of dielectric properties

2.2.1 Frequency-domain reflectometry

Equipment. DR measurements in the frequency range $0.2 \leq \nu/\text{GHz} \leq 20$ of most of the systems studied were performed with a Hewlett Packard HP85070B coaxial probe head connected to a HP8720D vector network analyzer (VNA)[104] at the institute of Prof. Glenn Hefter, Murdoch University, Perth, Australia. The probe head was placed inside a thermostated cell as described by Schrödle.[71]

Additionally, the new equipment at Regensburg, consisting of an Agilent E8364B VNA with an Agilent N4693A electronic calibration (ECal) module and dielectric probe kit 85070E, was used for measurements up to 50 GHz. The VNA controls the ECal module directly via an USB interface. The module corrects systematic errors, e.g. due to changes in the physical length of external and internal cables. Two probe heads are necessary to cover the full frequency range. A thermostated cell similar to the one described by Schrödle[71] was designed for mounting the high-temperature probe ($0.2 \leq \nu/\text{GHz} \leq 20$). The performance probe ($1 \leq \nu/\text{GHz} \leq 50$) was placed inside a newly developed thermostated cell, which allows effortless removal of the probe head for cleaning purposes (KIMTECH Wipes and acetone p.A.). Temperature was controlled by a Huber CC505 thermostat.

For both instruments, the temperature was monitored by a Pt-100 resistance with a precision of $\pm 0.02\,°\text{C}$ and an overall accuracy of $\pm 0.05\,°\text{C}$. An atmosphere of dry nitrogen was maintained during filling of the cells and the measurement itself.

Data acquisition and correction. The VNA determines the relative complex reflection coefficient, $\hat{\Gamma}_a$, at the probe head/sample interface, which is connected to the normalized aperture impedance of the probe head, $\hat{Y}$, via

$$\hat{\Gamma}_a = \frac{1-\hat{Y}}{1+\hat{Y}}. \tag{2.1}$$

The quantity of interest, i.e. the generalized complex permittivity of the sample, $\hat{\eta}$, was obtained by application of a simplified coaxial aperture opening model[105,106] and numerical solution of the equation

$$\hat{Y} = \frac{\mathrm{i}\hat{k}^2}{\pi \hat{k}_c \ln(D/d)} \left[\mathrm{i}\left(I_1 - \frac{\hat{k}^2 I_3}{2} + \frac{\hat{k}^4 I_5}{24} - \frac{\hat{k}^6 I_7}{720} + ...\right) + \left(I_2 \hat{k} - \frac{\hat{k}^3 I_4}{6} + \frac{\hat{k}^5 I_6}{120} - ...\right) \right], \tag{2.2}$$

where $\hat{k}_c = \omega\sqrt{\hat{\eta}_c \varepsilon_0 \mu_0}$ and $\hat{k} = \omega\sqrt{\hat{\eta}\varepsilon_0 \mu_0}$ are the propagation constants within the dielectric material of the coaxial probe head (*index* c) and within the sample, respectively. A theoretical

2.2. MEASUREMENT OF DIELECTRIC PROPERTIES

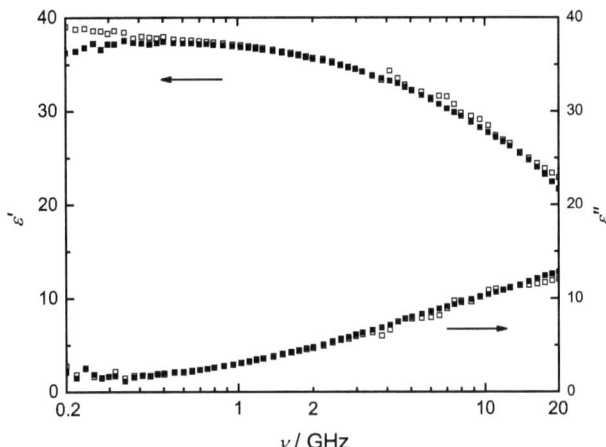

Figure 2.1: Dielectric permittivity, $\varepsilon'(\nu)$, and loss, $\varepsilon''(\nu)$, spectra (average of two independent calibrations) of [emim][EtSO$_4$] in acetonitrile ($w_{\text{[emim][EtSO}_4]} = 0.2982$). N,N-dimethylacetamide (■) and acetonitrile (□) were used as calibration materials.

approach yields the probe constants $I_1 \ldots I_{28}$.[106] The coaxial line outer and inner radii are indicated by D and d, respectively.

For the determination of $\hat{\Gamma}_\text{a}$, the instrument has to be calibrated to correct for signal distortions between the reference plane of the probe head and that of the VNA. A three point calibration, i.e. measurement of the uncorrected reflection coefficients, $\hat{\Gamma}$, of three standard materials was employed to calculate the complex, frequency dependent correction constants, $\hat{e}_\text{d}$, $\hat{e}_\text{r}$ and $\hat{e}_\text{s}$ according to the equation

$$\hat{\Gamma}_\text{a} = \frac{\hat{\Gamma} - \hat{e}_\text{d}}{\hat{e}_\text{s}(\hat{\Gamma} - \hat{e}_\text{d}) + \hat{e}_\text{r}}. \tag{2.3}$$

The standard materials include air (open), mercury (short) and a pure liquid (load) with accurately known dielectric properties. Mercury was purified according to the procedure described by Wölbl;[107] minor scum formation was removed by 'pin-hole' filtration (paper filter). The third calibration material has to be readily available in high purity and its dielectric properties should be close to those of the system studied. Benzonitrile (IL + IL mixtures), methanol

(IL + methanol mixtures) and N,N-dimethylacetamide (IL + acetonitrile mixtures) were used for measurements of the systems indicated in brackets. Use of acetonitrile as reference is not advisable as its dielectric properties result in oscillations in the detected signal (see Figure 2.1). Accordingly, DMA was adopted as the calibration material, because its static dielectric constant is close to that of acetonitrile at 25 °C (DMA: $\varepsilon = 38.25$[108] and AN: $\varepsilon = 35.84$[109]). Figure 2.1 shows the agreement obtained using independent calibrations with both solvents. The high-frequency oscillations are less pronounced when DMA is used as the standard. Therefore, DMA was chosen for calibration of IL + AN mixtures. At least two independent calibrations were performed to obtain a set of consistent spectra.

As mentioned, accurate data can only be obtained if the dielectric behavior of the reference and investigated samples are close. This condition is often not fulfilled. To correct for systematic errors, a complex Padé approximation,[110]

$$\hat{\varepsilon} = P_{n/m}\left[\hat{\varepsilon}^{\text{raw}}\right] = \frac{\hat{A}_0 + \hat{A}_1\hat{\varepsilon}^{\text{raw}} + ... + \hat{A}_n(\hat{\varepsilon}^{\text{raw}})^n}{1 + \hat{B}_1\hat{\varepsilon}^{\text{raw}} + ... + \hat{B}_m(\hat{\varepsilon}^{\text{raw}})^m}, \qquad (2.4)$$

was applied to calculate the corrected spectra, $\hat{\varepsilon}$, from the raw spectra, $\hat{\varepsilon}^{\text{raw}}$. The approximation constants, $\hat{A}_n(\omega)$ and $\hat{B}_m(\omega)$, were obtained by a complex fit algorithm[111] from a set of measurements of secondary standards, whose dielectric properties were close to the investigated systems. In this work, acetonitrile, benzonitrile, 1-butanol, N,N-dimethylacetamide and methanol together with their published relaxation parameters[108,109,112,113] were used for a $P_{1/1}$ correction. To avoid the artificial creation of oscillations upon the corrected spectra, acetonitrile was employed only when mixtures contained high amounts of that compound ($x_{\text{AN}} \gtrsim 0.8$).

For a crosscheck, raw and Padé corrected data were compared to interferometer measurements (IFM, see below), which are thought to be very reliable as they yield absolute values without the need for reference substances. These data are shown in Figure 2.2 together with the corresponding relative percentage deviations, $\delta\varepsilon' = 100 \cdot (\varepsilon'_{\text{IFM}} - \varepsilon')/\varepsilon'_{\text{IFM}}$ and $\delta\varepsilon'' = 100 \cdot (\varepsilon''_{\text{IFM}} - \varepsilon'')/\varepsilon''_{\text{IFM}}$, for the dispersion and loss curves, respectively. With the applied Padé correction, the deviations decreased considerably and were well within the experimental uncertainties of ca. $\pm 2\%$ in $\hat{\varepsilon}$.[38] Therefore, the procedure was applied to all VNA measurements.

2.2.2 Interferometry

Equipment. A set of double-beam interferometers was used for measurement of the dielectric properties at high GHz frequencies.[114] Electromagnetic waves are transmitted through rectangular waveguides of particular mechanical dimensions, each designated for a limited frequency range: $8.5 \leq \nu/\text{GHz} \leq 12$ (X-band, not used in this work), $13 \leq \nu/\text{GHz} \leq 17.5$ (Ku-band),

2.2. MEASUREMENT OF DIELECTRIC PROPERTIES

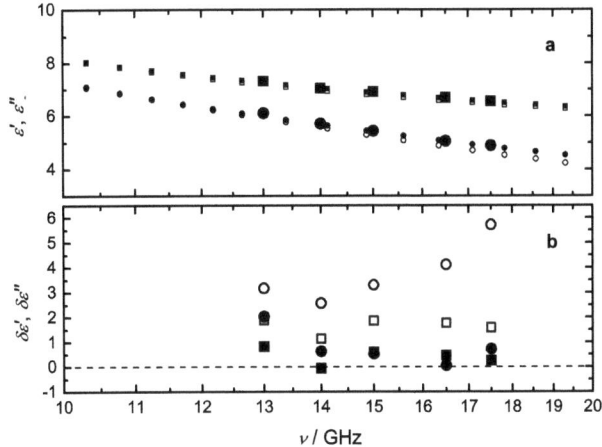

Figure 2.2: **(a)** Dielectric permittivity, $\varepsilon'(\nu)$ (squares), and loss, $\varepsilon''(\nu)$ (circles), data for a [bmim][BF$_4$] + methanol mixture at 25 °C ($w_{\text{[bmim][BF}_4\text{]}} = 0.9952$). Raw (small open symbols) and Padé corrected (small filled symbols) together with IFM data (large filled symbols) are shown. **(b)** Relative deviations of interpolated raw (open symbols) and Padé corrected (full symbols) data from IFM measurements.

$27 \leq \nu/\text{GHz} \leq 39$ (A-band) and $60 \leq \nu/\text{GHz} \leq 89$ (E-band). Signals are generated by frequency-stabilized sources and then split by a directional coupler to feed the measuring and reference branches. The sample is placed in a cell consisting of a piece of window-sealed waveguide and a gold-plated ceramic probe with variable, motor-controlled position. Variable precision phase shifters and attenuators are integrated in the branches. Signals are recombined by directional couplers and then measured by a precision attenuation receiver. Figure 2.3 shows a scheme of the E-band setup, which is similar to, within slight variations, the X-, Ku- and A-band instruments.*

The temperature was controlled by a Julabo FP 50 or a Lauda RK 20 thermostat and monitored

---

*To overcome the problem of time-demanding single frequency measurements, the X-, Ku- and A-band interferometers were connected to the VNA to enable continuous frequency scans.[40] This technique was used for test measurements, when the Agilent E8364B VNA became available.

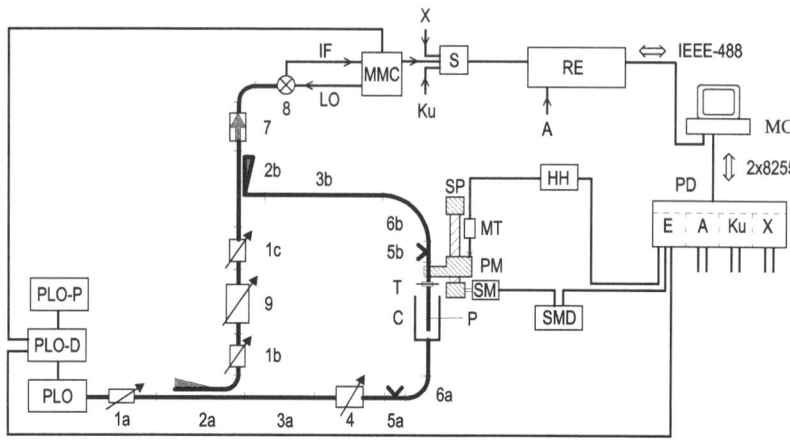

Figure 2.3: Block diagram of the E-band equipment:[114] **1a, b, c** represent variable attenuators; **2a, b** directional couplers; **3a, b** waveguide sections; **4** precision phase shifter; **5a, b** E/H tuners; **6a, b** flexible waveguides; **7** isolator; **8** harmonic mixer; **9** variable precision attenuator; **C** cell, **HH** bidirectional counter; **MC** microcomputer; **MMC** millimeter-wave to microwave converter; **MT** digital length gauge; **P** probe; **PD** parallel interface unit; **PLO** phase locked oscillators; **PLO-D** PLO-control unit; **PLO-P** PLO-power supply; **PM** probe mount; **RE** precision receiver; **S** electromechanical switch; **SM** stepping motor; **SMD** stepping motor control unit; **SP** spindle and spindle mount; **T** tapered transmission; double lines represent waveguides, thick lines semi-rigid microwave cables and normal lines symbolize data transfer connections (analog or digital).

by a Pt-100 resistance with a precision of $\pm 0.02\,^\circ\text{C}$ and an overall accuracy of $\pm 0.05\,^\circ\text{C}$. The filling of the cells and the measurements were performed under an atmosphere of dry nitrogen.

Data acquisition. After completion of all preliminary setup steps (e.g. filling of the cell, determination of the start position for the measurement), sample and reference beams are tuned to complete destructive interference at an appropriate probe position z'_0 with the help of attenuators and phase shifters. As the receiver detects only amplitudes, this procedure is necessary to obtain the phase information. The measurement itself includes the determination

2.2. MEASUREMENT OF DIELECTRIC PROPERTIES

of the relative attenuation as a function of the probe position.

The time-dependence of a harmonically oscillating field is described by

$$\hat{E}_1(t) = E_0 \exp(i\omega t), \qquad (2.5)$$

which can be assumed for an electromagnetic wave propagating through the reference beam. Provided that the interfering waves are fully destructive, their phases, φ, are shifted by

$$\Delta\varphi = (2n+1)\pi \quad \text{with } n \in \mathbb{Z}. \qquad (2.6)$$

Accordingly, the term π resulting from the condition of fully destructive interference is introduced, and it follows:

$$\hat{E}_2(t,x) = E_0 \exp(-\alpha x) \exp\left[i(\omega t + \pi - \beta x)\right], \qquad (2.7)$$

where α and $\beta = 2\pi/\lambda_\mathrm{m}$ are the absorption and phase coefficients; λ_m is the wavelength of the radiation within the sample and $x = z_0 - z_0'$ is the relative probe variation in the z-direction defined as the distance between the absolute optical path length of the sample, z_0, and the interference minimum, z_0'. The electric field reaching the receiver is the sum of fields $\hat{E}_1$ and $\hat{E}_2$,

$$\hat{E}(t,x) = \hat{E}_1(t) + \hat{E}_2(t,x) = \hat{E}_0 \exp(i\omega t)\left[1 + \exp(-\alpha x)\exp(i(\pi - \beta x))\right]. \qquad (2.8)$$

The power P of the signal detected by the receiver can be calculated via

$$P = \hat{E} \cdot \hat{E}^* = E_0^2 \cdot I(x), \qquad (2.9)$$

where E_0 is the amplitude of $\hat{E}$ and $I(x)$ is the interference function, which is defined as:

$$I(x) = [1 + \exp(-\alpha x)\exp(i(\pi - \beta x))] \cdot [1 + \exp(-\alpha x)\exp(i(\pi - \beta x))] \qquad (2.10)$$

$$= 1 + \exp(-2\alpha x) + \exp(-\alpha x) \cdot 2\cos(-\pi + \beta x). \qquad (2.11)$$

The signal-level measured by the receiver, $A(x)$, is commonly expressed in decibel (dB), i.e. the relative attenuation of the signal power on a logarithmic scale. It is defined via

$$A(x) = 10\lg\frac{P(x)}{P_\mathrm{ref}}. \qquad (2.12)$$

As P_ref is not known, $A(x)$ is normalized by A_0,

$$A_0 = 10\lg\frac{P_0}{P_\mathrm{ref}}, \qquad (2.13)$$

and it follows:

$$\begin{aligned}A_{\text{rel}}(x) &= A(x) - A_0 \\ &= 10\lg\frac{P(x)}{P_{\text{ref}}} - 10\lg\frac{P_0}{P_{\text{ref}}} \\ &= 10\lg\frac{P(x)}{P_0} \\ &= 10\lg\frac{E_0^2 \cdot I(x)}{E_0^2}.\end{aligned} \quad (2.14)$$

The recorded interference curve $A(z_0 - z_0')$ is fit by the expression[114]

$$\begin{aligned}A(z_0 - z_0') = A_0 &+ 10\lg\left\{1 + \exp\left[-2p\alpha_{\text{dB}}(z_0 - z_0')\right]\right. \\ &\left. - 2\cos\left(\frac{2\pi}{\lambda_{\text{m}}}(z_0 - z_0')\right) \cdot \exp\left[-2p\alpha_{\text{dB}}(z_0 - z_0')\right]\right\}.\end{aligned} \quad (2.15)$$

Eq. 2.15 yields the power attenuation coefficient, α_{dB} in dB/m, and λ_{m}; p is a conversion constant. These quantities are related to $\hat{\eta}(\nu)$ via:

$$\eta'(\nu) = \left(\frac{c_0}{\nu}\right)^2\left[\left(\frac{1}{\lambda_{\text{c},10}^{\text{vac}}}\right)^2 + \left(\frac{1}{\lambda_{\text{m}}(\nu)}\right)^2 - \left(\frac{\alpha_{\text{dB}}(\nu)}{2\pi}\right)^2\right] \quad \text{and} \quad (2.16)$$

$$\eta''(\nu) = \left(\frac{c_0}{\nu}\right)^2\frac{\alpha_{\text{dB}}(\nu)}{\pi\lambda_{\text{m}}(\nu)} \quad (2.17)$$

Here, $\lambda_{\text{c},10}^{\text{vac}}$ is the limiting vacuum frequency, a characteristic quantity for a particular waveguide.

2.2.3 THz time-domain spectroscopy

A THz time-domain spectrometer (THz-TDS) in transmission and reflection geometry was used in collaboration with Dr. Markus Walther and Dr. Andreas Thoman at the University of Freiburg.[§] A Ti:sapphire laser (Femtosource, Femtolasers Inc.; ~790 nm center wavelength) pumped by a 532 nm Nd:YVO$_4$ solid state laser (Verdi, Coherent Inc.) provided pulses of 12 fs duration with a repetition rate of 80 MHz and an average output power of 400 mW. For the generation and detection of THz radiation, the pulses were guided by a beam splitter and mirrors to photoconductive switches (Figure 2.4) made from semiconducting GaAs with DC biased metal striplines on the semiconductor. By focusing an ultrashort laser pulse in between the meltal striplines with a typical separation of $(30-80)\,\mu$m, charge carriers are created in

---

[§]We gratefully acknowledge the help of Dr. Andreas Thoman regarding the THz-TDS measurement of various samples.

2.2. MEASUREMENT OF DIELECTRIC PROPERTIES

the antenna. These are accelerated in the applied field, inducing a current $I(t)$ between the conducting lines, which gives rise to emission of electromagnetic radiation within the approximate range of 100 GHz to 5 THz. For the receiver, a low-temperature-grown GaAs substrate with an H-shaped metal stripline of $5\,\mu$m spacing was used. When the laser pulse coincides spatially and temporally with the THz field, a photocurrent is induced that is proportional to the incident electric field. The photocurrent was measured by means of a lock-in amplifier and a chopper wheel operating at ~330 Hz. By variation of the time delay between the laser pulse and the THz transient, the entire THz waveform was mapped. The transmission and reflection setups were kept in air-tight boxes purged with dry nitrogen to minimize absorption of the THz beam by water. A short introduction into the various setups will be given here; further details are given in refs. 115–117.

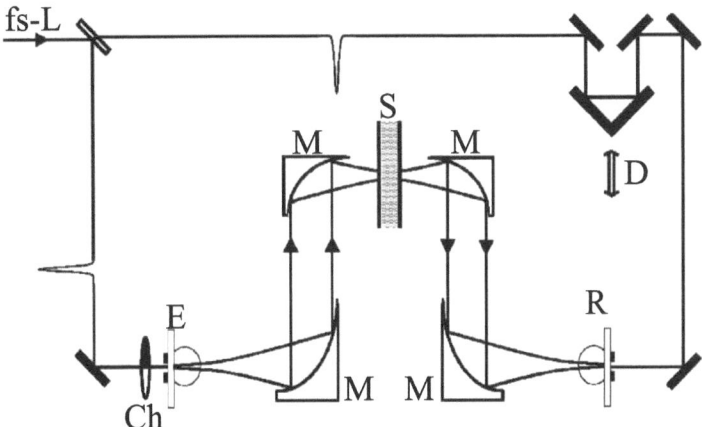

Figure 2.4: Schematic representation of the THz-TDS transmission setup consisting of **fs-L** fs-laser, **Ch** the chopper wheel, **M** parabolic mirrors, **S** liquid sample placed between two windows, **E** emitter antenna, **R** receiver antenna. **D** is a delay line in the gate beam.[71]

Transmission setup. The samples were placed in transmission cells with two parallel PTFE windows and an effective pathlength of ~1.5 mm as described by Schrödle.[71] This setup has several advantages, including ease of filling and temperature control, but covers only a limited frequency band ($\lesssim 2$ THz for the systems investigated) due to the declining signal-to-noise ratio at higher frequencies. Using thinner spacers would increase the bandwidth but at the cost of stronger multiple reflections. A cell with ~ 0.5 mm effective pathlength was designed and is currently being tested. The transmission setup is shown in Figure 2.4.

Temperature was controlled by a Julabo FP 50 thermostat and monitored using calibrated Pt-100 sensors (overall accuracy $\pm 0.05\,°C$).

The first step in a THz transmission measurement was the determination of the reference spectrum, $E_r(t)$, of the empty cell. This was followed by subsequent measurement of the sample spectrum, $E_s(t)$. The time-domain data so obtained were transformed into the frequency domain by the Fourier integral,

$$\hat{E}_{r,s}(\nu) = \frac{1}{2\pi}\int_{-\infty}^{+\infty} e^{-i 2\pi\nu t} E_{r,s}(t)\mathrm{d}t. \tag{2.18}$$

The refractive index, n_s, and absorption coefficient, α_s, of the sample were determined from the phase, $\Delta\phi$, and amplitude, A, of the ratio $\hat{R}(\nu)$ of the Fourier transforms of sample and reference pulses,

$$\hat{R}(\nu) = \hat{E}_s(\nu)/\hat{E}_r(\nu) = A(\nu)\cdot e^{i\Delta\phi(\nu)}, \tag{2.19}$$

via the relations:

$$n_s = n_{\text{air}} + c\,\Delta\phi/2\pi\nu d \quad \text{and} \tag{2.20}$$

$$\alpha_s = -2\ln(f_r \cdot A)/d, \tag{2.21}$$

where d is the thickness of the sample, $n_{\text{air}} = 1.00027$ is the refractive index of air and f_r is a correction coefficient accounting for reflection losses at the sample/PTFE interface.[71] The complex permittivity was calculated via[118]

$$\varepsilon'_s = n_s^2 - k_s^2 \quad \text{and} \tag{2.22}$$

$$\varepsilon''_s = 2 n_s k_s^2, \tag{2.23}$$

with $k_s = c_0\alpha_s/4\pi\nu$.

Reflection setup. As mentioned above, transmission measurements were limited in the accessible frequency range. To circumvent these limitations, a reflection setup was used to measure highly absorbing liquids up to ~ 3 THz. The setup is shown in Figure 2.5. Technical principles are the same in transmission and reflection measurements, but now only the reflected part of the signal is of interest. The THz pulse generated at the emitter was partly transmitted through

2.2. MEASUREMENT OF DIELECTRIC PROPERTIES

a silicon wafer (beam splitter). This remaining pulse was focused on the sample (or reference), which was placed on a silicon window. The THz pulse was reflected at the sample/silicon (reference/silicon) surface back to the silicon wafer and subsequently detected by the receiver. The reflection at the front side of the silicon window was used to check the absolute positions of the silicon window in the reference and sample measurements. In that way, potential positioning errors could be corrected.

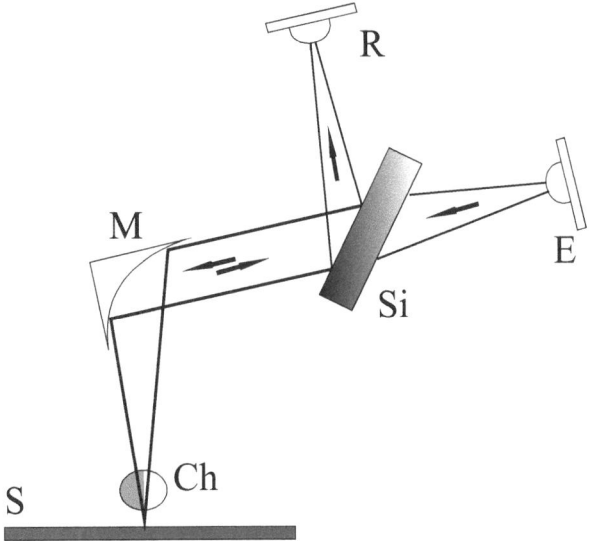

Figure 2.5: Schematic representation of the THz-TDS reflection setup consisting of **S** liquid sample, **Ch** the chopper wheel, **M** parabolic mirror, **Si** silicon wafer, **E** emitter antenna, **R** receiver antenna.

At the beginning of this thesis work, measurements were performed at room temperature. Later, the setup was equipped with a cell, that allowed connection to a thermostat. Temperature was determined with calibrated Pt-100 sensors (overall accuracy $\pm 0.05\,°C$).

The field strength of the reflected fraction, E^{refl}, of an incoming THz pulse, E^{inc}, is given by Fresnel's equation,

$$E_{\text{s,r}}^{\text{refl}} = E^{\text{inc}} \frac{\hat{n}_{\text{Si}} - \hat{n}_{\text{s,r}}}{\hat{n}_{\text{Si}} + \hat{n}_{\text{s,r}}}, \qquad (2.24)$$

where $\hat{n}_\text{Si}$ is the complex index of refraction of the silicon window and $\hat{n}_\text{s,r}$ of the sample or reference, respectively. Air was used as reference material. Assuming $\hat{n}_\text{Si} = n_\text{Si}(= 3.42)$ and $\hat{n}_\text{r} = n_\text{r}(= 1.00027)$,[71,115] the relation

$$\frac{E_\text{s}^\text{refl}}{E_\text{r}^\text{refl}} = \frac{n_\text{Si} - \hat{n}_\text{s}}{n_\text{Si} + \hat{n}_\text{s}} \cdot \frac{n_\text{r} + n_\text{Si}}{n_\text{Si} - n_\text{r}} \qquad (2.25)$$

yields the complex index of refraction of the sample, $\hat{n}_\text{s} = n_\text{s} - ik_\text{s}$, which is transformed to $\varepsilon'(\nu)$ and $\varepsilon''(\nu)$ via Eqs. 2.22 and 2.23.

2.3 Supplementary measurements

2.3.1 Density

To calculate molar concentrations, c, the densities, ρ, of the investigated systems were determined with a vibrating tube density meter (Anton Paar DMA 60, DMA 601HT) according to the method of Kratky *et al.*[119] The temperature was kept constant to $\pm 0.003\,°\mathrm{C}$ with an overall accuracy of $\pm 0.02\,°\mathrm{C}$ using a Braun Thermomix 1480 thermostat in combination with a thermostated heat sink (Lauda RK 20). The instrument was calibrated with degassed water (Millipore MILLI-Q) and purified nitrogen at atmospheric pressure, assuming densities from standard sources.[120] The precision of the measurements was $\pm 0.001\,\mathrm{g\,L^{-1}}$. Taking into account all sources of error (calibration, measurement, purity of materials), the overall uncertainty of ρ is estimated to be within $\pm 0.05\,\mathrm{g\,L^{-1}}$.

In addition, the densities of selected [emim][EtSO$_4$] + AN mixtures were measured with a density meter DMA 5000 M (Anton Paar GmbH, Graz, Austria).† The temperature was kept constant to $\pm 0.001\,°\mathrm{C}$ with an overall accuracy of $\pm 0.01\,°\mathrm{C}$ with a built-in thermoelectric temperature control. The precision of the measurements was $\pm 0.001\,\mathrm{g\,L^{-1}}$. The accuracy stated by the manufacturer is $\pm 0.005\,\mathrm{g\,L^{-1}}$, although real errors are certainly higher.

The data obtained for selected [emim][EtSO$_4$] + AN mixtures using the DMA 5000 M density meter are given in Appendix A.1.

2.3.2 Conductivity

Electrical conductivities, κ, were determined with the equipment described by Barthel and co-workers,[121,122] consisting of a home-built precision thermostat stable to $< 0.003\,\mathrm{K}$ in combination with a thermostated heat sink (Lauda Kryomat K 90 SW). Additionally, a new homemade precision thermostat was set up in combination with a thermostated heat sink (Julabo FP40, flow rate tunable via a needle valve) for measurements at high temperatures. Within slight variations, the setup corresponds to that shown in Figure 1 of ref. 123. High-temperature stable materials were used throughout. This thermostat was stable to $< 0.003\,°\mathrm{C}$ over the entire temperature range. With the currently used silicon oil (M50), it is limited to $\lesssim 215\,°\mathrm{C}$. The resistance of the temperature bridge, R_t, was calibrated with the use of a NIST-traceable Pt sensor and bridge (ASL).

---

†We thank C. Schöggl-Wagner and T. Feischl, Anton Paar GmbH, Graz, Austria, for performing the measurements.

One set of three-electrode and two sets of two-electrode capillary cells were used depending on the values of κ. The cells with cell constants C in the range of $(25 \text{ to } 470)\,\text{m}^{-1}$, $(12 \text{ to } 46)\,\text{cm}^{-1}$ and $(25 \text{ to } 360)\,\text{cm}^{-1}$, respectively, were calibrated with aqueous KCl.[124]
The cell resistance, $R(\nu)$, was measured with a manually balanced high-precision conductivity bridge as a function of AC frequency, ν, between 480 Hz and 10 kHz. To eliminate electrode polarization, the measured resistances were extrapolated to infinite frequency, $R_\infty = \lim_{\nu \to \infty} R(\nu)$, using the empirical function $R(\nu) = R_\infty + A/\nu^a$; the parameter A was cell-specific, and the exponent a was found to be in the range $0.5 \lesssim a \lesssim 1$.[125] The conductivity was obtained as $\kappa = C/R_\infty$.

The uncertainty in temperature was $\pm 0.01\,°\text{C}$. Repeat measurements of selected samples with different cells agreed within $\pm 0.5\,\%$. Thus, this value may be taken as an estimate for the relative uncertainty of the obtained values.[22,97] The effect of temperature on the cell constants C was well below the stated accuracy.[97]

The setup described was used to determine precise conductivities of binary IL + AN mixtures at $25\,°C$[22] and of neat imidazolium-based ILs as a function of temperature.[97]

2.3.3 Viscosity

Viscosities, η, of [emim][BF$_4$]#2 + [emim][DCA] mixtures (Section 4) were measured under an argon atmosphere at $(25 \pm 1)\,°\text{C}$ using a CVO 120 high-resolution rotational viscometer (Bohlin Instruments, UK) and a cone of $4°$ slope with 40 mm diameter. The instrument was calibrated by calculation of a correction factor, $f_\text{corr} = \overline{\eta}_\text{Lit}/\eta$, using averaged published data, $\overline{\eta}_\text{Lit}$, for [emim][BF$_4$].[126–128] A similar value for f_corr, albeit with a much larger uncertainty, was obtained for neat [emim][DCA] but was not considered further because the published viscosities scattered considerably. The corrected viscosities were than obtained as $\eta_\text{corr} = f_\text{corr} \cdot \eta$. Repeat measurements indicate a precision of $\pm 2\,\%$. This value may be taken as an estimate for the uncertainty of the measurement.

Additional measurements of selected [emim][EtSO$_4$] + AN mixtures were performed using a AMVn automated micro falling ball viscometer (Anton Paar GmbH, Graz, Austria) with built-in thermoelectric temperature control ($0.01\,°\text{C}$ resolution; $< 0.05\,°\text{C}$ accuracy).[†] The instrument was equipped with capillaries of varying diameter, $d_\text{cap} = 1.6$ or $1.8\,\text{mm}$, and fitting balls calibrated by the supplier. The diameter of the capillary was chosen with respect to an optimized rolling time of the ball. With these two capillaries measurements in the range $0.3 \leq \eta/\text{mPa s} \leq 70$ are realizable. The precision stated by the manufacturer is better than $0.5\,\%$.

The data obtained for selected [emim][EtSO$_4$] + AN mixtures using the AMVn viscometer are listed in Appendix A.1.

2.3.4 Refractive indices

Refractive indices, n, of [emim][EtSO$_4$] + AN mixtures were measured at 25 °C using an Abbemat WR MW automatic digital refractometer (Anton Paar GmbH, Graz, Austria), equipped with a custom-made, variable light source made from LEDs (tunable to 437.0, 488.1, 515.0, 531.8, 589.2 and 632.2 nm).[‡] The instrument measured the critical angle of total reflection by shadowline detection with a CCD array. The temperature was adjusted with a built-in Peltier thermostat, with a stated stability of ±0.002 °C and accuracy of ±0.03 °C. As specified by the manufacturer, the resolution and overall accuracy of the measurement were $1 \cdot 10^{-6}$ and $4 \cdot 10^{-5}$, respectively. Repeat measurements were well within the latter range.

The data obtained for [emim][EtSO$_4$] + AN mixtures are listed in Appendix A.1.

2.4 Raman spectroscopy

Raman spectra of [emim][EtSO$_4$] + AN mixtures were measured in the frequency range $50 \leq \bar{\nu}/\text{cm}^{-1} \leq 4000$ at ambient temperature ($\theta \approx 22\,°\text{C}$) using the equipment consisting of a Bruker IFS 66 FT-IR spectrometer with a FR 106 FT-Raman accessory kit and a D 418-S high sensitivity detector.[‡] The power of the Nd:YAG laser working at 1064 nm was ∼100 mW. Samples were filled into 10 mm quartz cells under an atmosphere of nitrogen. The data obtained was fit using the PeakFit program (version v4.11, Systat Software GmbH, Erkrath, Germany).

[‡]We thank Prof. Augustinus Asenbaum and Dr. Christian Pruner for giving me access to their equipment and for valuable support regarding the Raman measurements.

Chapter 3

Neat Components

3.1 Ionic liquids

Parts of the material presented in this chapter were published in:

Stoppa, A.; Hunger, J.; Thoman, A.; Helm, H.; Hefter, G.; Buchner, R. *'Interactions and Dynamics in Ionic Liquids.'* J. Phys. Chem. B **2008**, *112*, 4854.

Being responsive to dipolar species, the frequency-dependent dielectric function, $\hat{\varepsilon}(\nu)$, provides direct access to molecular-level interactions and dynamics over very wide ranges of molecular size and time scales.[42] Thus, DR spectroscopy should be able to give a comprehensive picture of the liquid-state dynamics of ILs. This is important for understanding solvation phenomena in ILs, as the time-dependent response of the solvent to solute-induced perturbations controls ultrafast reactions, especially charge-transfer processes.[129–131] It is well-known that solvation dynamics in polar liquids is essentially determined by the frequency-dependent dielectric function of the solvent and can be described within the framework of dielectric continuum models.[132] However, recent applications of this approach to solvation dynamics in ILs were not successful[129,130] because a considerable part of the short-time dynamics was missed by the continuum models used. At least in part this was a consequence of the limited frequency coverage of the published dielectric spectra of ILs. Work described in this thesis aimed to rectify this situation by combining complex permittivity data obtained with VNA, IFM and THz-TDS measurements.[36] While this approach showed that the accuracy of the continuum models could be improved by consideration of the broadband measurements described here (and published in ref. 36),[133] only combination of DR and far-infrared (FIR) measurements by Hunger[39,40] fully closed the gap on the high-frequency wing of the DR spectrum. Despite ongoing work in the field of solvation dynamics,[134] the recently published data[39] was not yet implemented in the literature.

This chapter will not give a detailed presentation of the most recent work in the field of neat ILs. This was one of the main subjects of the PhD thesis of J. Hunger.[40] It is rather intended to highlight some DR results, as their knowledge will be essential for the analysis of DR spectra of IL + IL (Section 4) and IL + solvent mixtures (Section 5).

Broadband dielectric properties. Figure 3.1 (taken from ref. 40) shows the experimental DR spectrum of a representative IL over a broad frequency range, which was covered by combination of data obtained from VNA, IFM, THz-TDS and FIR measurements. Various imidazolium-based ILs have been studied, including some with anions having zero (BF_4^-, PF_6^-) or small (DCA^-, $CF_3CO_2^-$) dipole moments. All of the presented spectra show broad peaks in $\varepsilon''(\nu)$ at low GHz and THz frequencies, connected by a reasonably distinct plateau in the 100 GHz region.

Analysis based on only DR spectra revealed the presence of five modes (Figure 3.1): a low frequency CC^m mode, a D^m mode in the ~ 100 GHz region and three DHO modes at THz frequencies. However, concerted analysis, i.e. the application of the same fit model to DR and optical herterodyne-detected Raman-induced Kerr effect (OHD-RIKE) spectra shed more light onto the nature of these liquids.[39] In addition to the fit of DR spectra alone,[40] a D relaxation at ~ 1 GHz could be resolved, as it is characterized by a large OHD-RIKE feature, but small DR amplitude. Its origin is probably due to breathing modes of high-symmetry clusters. It was explained by the presence of a mesoscopic structure in imidazolium-based ILs, which is formed by aggregation of nonpolar (alkyl groups) and polar (cation head groups and anions) domains, whose existence has been suggested by MD simulations[135,136] and experimentally shown by several techniques.[137–141] The CC^m and D^m modes, which are undoubtly both present in the DR and OHD-RIKE spectra, were explained by large-angle jump reorientation of the dipolar cations for the former mode, whereas the origin of the latter is still not clear. Contributions from intermolecular vibrations and rotational-translational cross-correlations seem to be plausible. Due to the high signal-to-noise ratio of the OHD-RIKE data, up to six modes could be resolved in the THz region, which are mainly due to librations or inter- and intramolecular vibrations. The exact number of modes is speculative at present due to the high degree of overlap, and thus an assignment to molecular level motions is not yet possible. Note that for the present anions as well as for the NTf_2^- anion[38] no contribution of these species except intramolecular vibrations, could be resolved in the analysis of DR and OHD-RIKE spectra. The only exception within the up-to-now studied ILs is the strongly polar $EtSO_4^-$ anion, where an additional D mode arising from anion reorientation appears at ~ 10 GHz in the DR spectra of [emim][$EtSO_4$].[37]

3.1. IONIC LIQUIDS

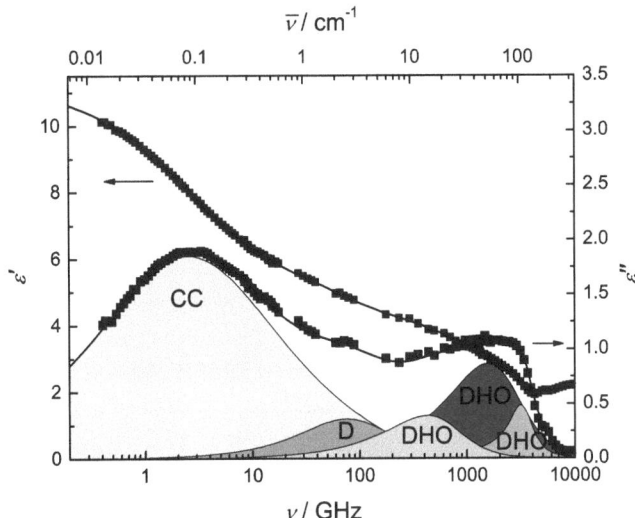

Figure 3.1: Dielectric permittivity, $\varepsilon'(\nu)$, and loss, $\varepsilon''(\nu)$, spectra of [bmim][DCA] at 25 °C (taken from ref. 40). Shaded areas indicate the contributions of the individual processes to $\varepsilon''(\nu)$.

Spectra at $\nu \leq 89\,\text{GHz}$. Most DR spectra recorded for neat ILs and IL + solvent mixtures to date have been restricted to $\nu \leq 89\,\text{GHz}$.[35,37,38,48] DR spectra at these frequencies for neat ILs can be fit to a CC + D model.[38] The limitations with respect to the frequency range covered imply the following restrictions: firstly, due to its low amplitude, the lowest-frequency D relaxation (breathing mode) cannot be resolved on the basis of DR spectra alone. Moreover, Figure 3.1 shows that up to three distinct high-frequency relaxations contribute in the 100 GHz region, i.e. the D^{m} mode and up to two DHO modes. Consequently, the highest-frequency D relaxation fit to spectra at $\nu \leq 89\,\text{GHz}$ is a superposition of various modes, which cannot be resolved as their maxima are located outside the considered frequency range. However, Hunger[40] could show from analysis of broadband DR spectra ($0.2 \leq \nu/\text{GHz} \leq 10000$) that microwave modes are still well characterized when spectra were limited to $\nu \leq 89\,\text{GHz}$.

3.2 Acetonitrile

DR spectra at $\nu \leq 89\,\text{GHz}$. The previously-performed temperature dependent DR study of neat AN (see ref. 109) was extended to lower temperatures, hence the range $-5 \leq t/°\text{C} \leq 35$ is covered now. The present DR spectra covering the frequency range $0.1 \leq \nu/\text{GHz} \leq 89$ were again best described by a single D equation. The parameters obtained are summarized in Table 3.1.

Table 3.1: Fit Parameters of Eq. 1.61 for the Observed Dielectric Spectra of Neat AN as Function of Temperature, θ, Assuming the D Model: Static Permittivities, ε; Relaxation Times, τ_1; Infinite Frequency Permittivity, ε_∞; and Reduced Error Function of the Overall Fit, χ_r^2.

$\theta/°\text{C}$	ε	τ_1/ps	ε_∞	χ_r^2
-5	40.89	4.58	3.73	0.043
0	40.16	4.32	3.67	0.021
5^a	39.08	4.06	3.56	0.028
15^a	37.41	3.62	3.38	0.021
25^a	35.84	3.32	3.33	0.006
35^a	34.51	3.06	3.38	0.019

^{a} taken from ref. 109.

The dispersion amplitude of liquid AN has often been interpreted in terms of the Kirkwood factor, g_K, and Eq. 1.67 gives a connection to experimentally accessible parameters. The absolute value of g_K is dependent on the choice of ε_∞; the values given here (Table 3.2) were calculated using FIR data published by Ohba et al.[142]

In concordance with other experimental (see ref. 143 and literature cited therein) and theoretical[144–149] techniques, a value of $g_\text{K} < 1$ was found from the present temperature dependent study. This indicates a preferentially antiparallel orientation of the molecular dipoles. Due to increased thermal motions of the molecules, a rise in temperature leads to an increase of g_K, which means the tendency towards statistical arrangement of the dipoles is enhanced.

The dynamic properties of AN can be analyzed in terms of microscopic relaxation times, τ_1', calculated according to Eq. 1.80 and given in Table 3.2. The decrease of τ_1' with respect to temperature is related to the decrease of viscosity[54] according to the SED model (Eq. 1.77).

3.2. ACETONITRILE

Table 3.2: Kirkwood Factors, g_K, and Microscopic Relaxation Times, τ_1', of Neat AN as Function of Temperature, θ.

$\theta/°C$	g_K	τ_1'
-5	0.765	3.12
0	0.773	2.94
5[a]	0.774	2.77
15[a]	0.783	2.47
25[a]	0.791	2.28
35[a]	0.805	2.10

[a] taken from ref. 109.

The values of τ_1' are well described by the linear equation

$$\tau_1'/\text{ps} = 0.6991 + 1376 \cdot \frac{\eta/\text{mPa} \cdot \text{s}}{T/\text{K}} \quad (\sigma = 0.3\,\text{ps}) \quad (3.1)$$

From Eqs. 1.77 & 1.79 and the slope of Eq. 3.1, it follows that $V_{\text{eff}} = 6.31\,\text{Å}^3$ for AN. The value of V_{eff} is connected to the molecular volume of AN ($V_m = 43.9\,\text{Å}^3$)[85] via the friction parameter, C, and the shape factor, f. The latter was calculated from Eq. 1.78 and the geometrical parameters of AN ($a = 2.90\,\text{Å}$; $b = 1.90\,\text{Å}$),[150,151] yielding $f = 1.208$. Consequently, one obtains $C = 0.119 \pm 0.002$, which agrees well with the theoretically calculated value, assuming *slip* boundary conditions, of $C_{\text{slip}} = 1 - f^{-2/3} = 0.118$.

This shows, that the SED model is fully applicable to neat AN. It indicates rotational diffusion of the molecular dipoles under *slip* boundary conditions as the mechanism for dielectric relaxation. Similar results have been deduced from DR studies of electrolyte solutions in AN.[85]

Among the various models available,[109] an Eyring equation was used to describe the temperature dependence of the relaxation times τ_1 of AN. This model was chosen because the small number of data points does not justify the application of fits with more adjustable parameters. A linear fit of $\ln \tau_1$ vs. $1/T$ (Eq. 1.101) yields $\Delta S^{\neq} = (-9.8 \pm 0.6)\,\text{J K}^{-1}\,\text{mol}^{-1}$ and $\Delta H^{\neq} = (4.6 \pm 0.2)\,\text{kJ mol}^{-1}$. The latter is similar to the values calculated from published DR data[108,112] of the dipolar aprotic liquids BN ($10.8\,\text{kJ mol}^{-1}$), PC ($16.5\,\text{kJ mol}^{-1}$) and DMA ($8.7\,\text{kJ mol}^{-1}$). The values of the entropies of activation, $\Delta S^{\neq}$, are $-8.4\,\text{J K}^{-1}\,\text{mol}^{-1}$ for BN, $9.1\,\text{J K}^{-1}\,\text{mol}^{-1}$ for PC and $-9.2\,\text{J K}^{-1}\,\text{mol}^{-1}$ for DMA. Negative values of $\Delta S^{\neq}$ may be interpreted in terms of collective reorientational motions of molecules around a central molecule, as was found in molecular dynamics simulations for neat AN,[152,153] but the results should not be over-interpreted as the experimental data is limited with respect to the temperature range investigated.

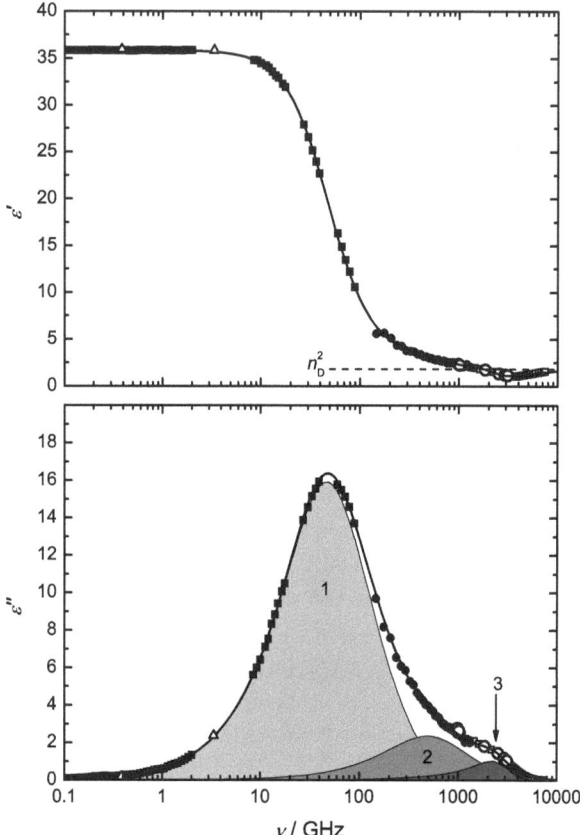

Figure 3.2: Dielectric permittivity, $\varepsilon'(\nu)$, and loss, $\varepsilon''(\nu)$, of AN at 25 °C: ●, experimental data of this work; data taken from: ■, ref. 109; △, ref. 154; ○, ref. 155; □, ref. 142. Lines represent fit with the D^m + DHO + DHO model, shaded areas indicate the contributions of the D^m (1) and the two DHO (2 & 3) modes. The square of the refractive index of AN at 589 nm, $n_D^2 = 1.80$,[120] is indicated.

3.2. ACETONITRILE

Extended frequency range. The maximum of the dielectric loss of acetonitrile is located at $\sim 48\,\text{GHz}$ at $25\,°\text{C}$, that is close to the upper limit of the instruments currently available in Regensburg. To gain a molecular-level understanding of solvent dynamics at shorter times data up to far-infrared frequencies (FIR, $\bar{\nu} \lesssim 300\,\text{cm}^{-1}$, $\nu \lesssim 10\,\text{THz}$) have to be available. Particularly in the late 1980's, significant attempts have been taken to make the experimentally demanding FIR region accessible. These included quite a few publications reporting the FIR optical properties of neat AN,[142,154–160] and selected literature data[142,154,155] were used to extend the available MW spectrum (at $\nu \leq 89\,\text{GHz}$)[109] at $25\,°\text{C}$. The then still existing gap in the $0.1 - 1\,\text{THz}$ region was bridged with a THz-TDS instrument operating in reflection geometry. The complex permittivity spectrum for AN covering the full range from MW to FIR frequencies is shown in Figure 3.2.

At high frequencies, say $\nu \gtrsim 200\,\text{GHz}$, the spectrum of AN (Figure 3.2) deviates considerably from the shape expected for a simple exponential relaxation. Among all various models tested, a combination of three modes described the present spectrum best (i.e. yielded the lowest value of χ_r^2): the main dispersion step centered at $\sim 48\,\text{GHz}$ was modelled by a D^m equation, and at higher frequencies two DHO modes peaking at $\sim 490\,\text{GHz}$ and $\sim 2.1\,\text{THz}$ could be resolved (D^m + DHO + DHO model). A Debye equation was used in previous studies[85,109,143] to describe the main relaxation in neat AN and was also tested in this work. However, for the present broadband spectrum a D^m equation was used to avoid unphysical FIR contributions of the Debye mode. Figure 3.2 shows the overall fit together with the resolved relaxations. The current model provides an exzellent fit of the present data at $25\,°\text{C}$. The parameters of the D^m + DHO + DHO fit are given in Table 3.3. Compared to the parameters obtained from fitting $\hat{\varepsilon}(\nu)$ up to $89\,\text{GHz}$ (Table 3.1), the amplitude of the main relaxation, S_1, is smaller due to the obvious overlap of process 2 (Figure 3.2), but ε and τ_1 agree perfectly. Hence, the parameters given in Table 3.1 give a reliable characterization of the microwave mode, even without resolving the high-frequency processes.

The main dispersion step, which is better described by a D^m equation, has been ascribed to rotational diffusion of molecular dipoles (see above). At higher frequencies, libration modes, which are known to be well described by DHO equations,[161] may occur. From the symmetry of AN, only one libration mode would be expected. Librational motions have been shown to contribute significantly to FIR[162] and OHD-RIKE spectra[163] of AN, although further spectral contributions may exist. Molecular-dynamics simulations have indeed shown,[164,165] that the FIR absorption of AN is due to various molecular motions: beside contributions from induced moments and cross-correlations, intermolecular vibrations mainly contribute in the $10-30\,\text{cm}^{-1}$ ($0.3 - 0.9\,\text{THz}$) region, whereas librations dominate at $30 - 60\,\text{cm}^{-1}$ ($0.9 - 1.8\,\text{THz}$). Thus,

Table 3.3: Fit Parameters of Eq. 1.61 for the Observed Dielectric Spectrum ($0.1 \leq \nu/\text{GHz} \leq 7500$) of Neat AN at 25°C Assuming a D^m + DHO + DHO Model: Static Permittivity, ε; Relaxation Time, τ_1; Inertial Rise Constant, γ_{lib}; Amplitudes, S_j; Resonance Frequencies, $\nu_{0,j}$; Damping Constants, γ_j; of Process j; Infinite Frequency Permittivity, ε_∞.[a]

ε	τ_1/ps	γ_{lib}/THz	S_1	$\nu_{0,2}$/THz	γ_2/THz	S_2	$\nu_{0,3}$/THz	γ_3/THz	S_3	ε_∞
35.83	3.32	2.5[b]	29.1	1.12	2.90	4.02	2.53	2.73	1.02	1.71

[a] $\chi_r^2 = 1.1 \cdot 10^{-2}$; [b] Parameter fixed during fitting procedure.

a definite assignment of the DHO modes to discrete molecular motions is questionable, but one may speculate that intermolecular vibrations and librations are the major contributors to processes 2 & 3, respectively, in Figure 3.2.

3.3 Methanol

The DR spectrum of MeOH in the range $0.05 \leq \nu/\text{GHz} \leq 5000$ at 25 °C was published and analyzed in terms of dynamic features by Fukasawa et al.[166] The authors used a D + D + D + DHO + DHO model to describe the full spectrum, but did not give all the corresponding parameters. For the quantitative analysis of IL + MeOH spectra (Section 5.2.2), the amplitudes of the modes in neat MeOH are required and thus, the published data were fit applying the same model. The parameters obtained are listed in Table 3.4 and the DR spectrum is shown in Figure 3.3.

As for other H-bonded liquids, the main relaxation step of MeOH was related to the cooperative H-bond relaxation of oligomeric chains.[166] The so-called *switch-over* mechanism was used to describe the cooperative rearrangement of all dipoles forming an oligomeric chain, which is induced by the break-up of an H-bond at the end of that chain.[113] The second (next highest frequency) relaxation process was assigned to the reorientation of individual dipoles.[166] The origin of process 3 is not yet fully understood. Comparison with Raman spectra indicated that it is not related to diffusive motion; it may reflect other microscopic dynamics, like the partial rotation of H-bond acceptors within the oligomeric chains.[90] The corresponding resonant THz contributions, peaking at 55 cm^{-1} (1.65 THz) and 130 cm^{-1} (3.90 THz),[167–169] have also been observed in the FIR spectra of MeOH. It has been suggested, that the lower-frequency mode may be ascribed to a vibration arising from the fluctuation of alkyl groups within the H-bond network,[168] whereas the second mode is most probably due to the intermolecular H-bond stretching vibration.

Table 3.4: Fit Parameters of Equation 1.61 for the Published Dielectric Spectrum ($0.05 \leq \nu/\text{GHz} \leq 5000$) of Neat MeOH at 25°C Assuming the D + D + D + DHO + DHO Model:[166] Static Permittivity, ε; Relaxation Times, τ_j; Amplitudes, S_j; Resonance Frequencies, $\nu_{0,j}$; Damping Constants, γ_j; Infinite Frequency Permittivity, ε_∞.[a]

ε	τ_1/ps	S_1	τ_2/ps	S_2	τ_3/ps	S_3	...
32.49	52.2	26.3	8.23	1.27	1.05	1.93	
...	$\nu_{0,4}/\text{THz}$	γ_4/THz	S_4	$\nu_{0,5}/\text{THz}$	γ_5/THz	S_5	ε_∞
	2.68	7.04	0.976	3.63	2.11	0.151	1.82

[a] $\chi_r^2 = 3.3 \cdot 10^{-3}$.

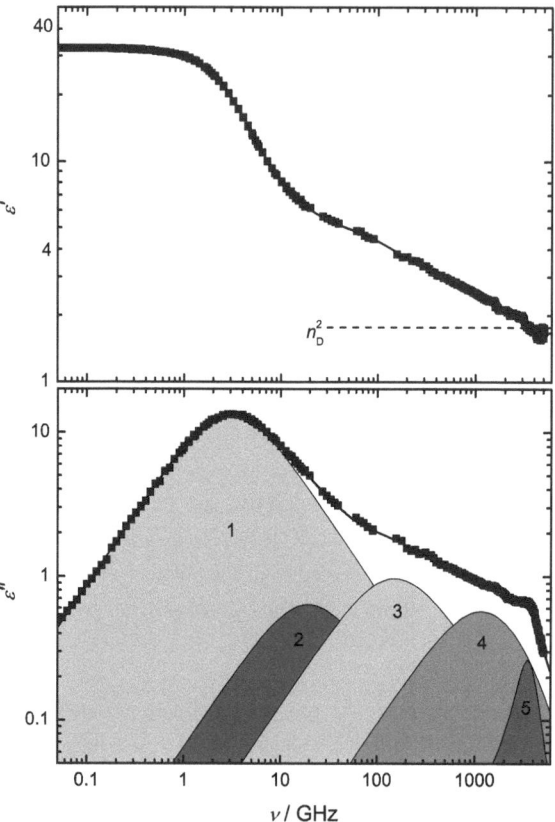

Figure 3.3: Dielectric permittivity, $\varepsilon'(\nu)$, and loss, $\varepsilon''(\nu)$, of MeOH at 25 °C (data taken from ref. 166). Lines represent fit with the D + D + D + DHO + DHO model, shaded areas indicate the contributions of the three D (1, 2 & 3) and the two DHO (4 & 5) modes. The square of the refractive index of MeOH at 589 nm, $n_D^2 = 1.77$,[120] is indicated. A logarithmic presentation of $\hat{\varepsilon}$ was chosen to allow proper comparison with Figure 3 of ref. 166.

Chapter 4

IL + IL mixtures

The material presented in this chapter was published in:

Stoppa, A.; Hefter, G.; Buchner, R. *'How Ideal are Binary Mixtures of Room-Temperature Ionic Liquids?'* J. Mol. Liq. **2010**, *153*, 46.

as part of the Special Issue: *'Understanding solvation from liquid to supercritical conditions', selected papers on molecular liquids presented at the EMLG/JMLG 2008 annual meeting, 31 August - 4 September 2008, Lisboa, Portugal.*

This chapter presents data for the densities, conductivities, viscosities and dielectric properties for [emim][BF$_4$] + [emim][DCA] mixtures (Table 4.1). The higher quality batch [emim][BF$_4$]#2 was used for the density, conductivity and viscosity measurements, whereas DR spectra were recorded using the commercial batch [emim][BF$_4$]#1 (Section 2.1.2). Fortunately, the DR spectra are relatively insensitive to minor impurities. The data have been analyzed in terms of their 'excess' properties to infer the effects of anion exchange on the intermolecular interactions in binary mixtures of imidazolium ILs.

4.1 Fit model

DR spectra of the binary mixtures were measured in the frequency range $0.2 \leq \nu/\text{GHz} \leq 20$ (Figure 4.1). As the fit parameters describing the main relaxation observed are negligibly affected by the frequency range covered (see below), the measurements were not extended to 89 GHz.

Consistent with previous investigations of neat ILs over the wider frequency range of $0.2 \lesssim \nu/\text{GHz} \leq 89$,[38] the best fit was found by adopting a Cole-Cole relaxation centered around 3 GHz

Table 4.1: Mole Fractions, x_2, and Molar Concentrations, c_i, of Binary [emim][BF$_4$] (**1**) + [emim][DCA] (**2**) Mixtures, Together with Measured Densities, ρ, Electrical Conductivities, κ, and Corrected Viscosities, η_{corr} (Section 2.3.3) at 25 °C.

x_2	c_2 mol L^{-1}	c_1 mol L^{-1}	ρ g L^{-1}	κ S m^{-1}	$10^3 \eta_{\text{corr}}$ Pa s
0	0	6.468	1280.38	1.553	36.8
0.05527	0.3566	6.095	1269.75	1.636	34.7
0.2144	1.373	5.034	1239.88	1.874	31.4
0.3746	2.384	3.981	1210.49	2.10	27.9
0.5260	3.328	2.999	1183.40	2.30	23.0
0.6782	4.266	2.025	1156.79	2.49	21.4
0.7836	4.910	1.356	1138.51	2.61	19.7
0.9552	5.949	0.2788	1109.35	2.79	17.2
1	6.218	0	1101.87	2.83	17.4

and an additional Debye relaxation at higher frequencies, i.e. the CC + D model. Fits assuming a single process, described for instance by a HN equation (Eq. 1.56), yielded significantly larger χ_r^2. Since for the CC + D model the loss peak of the fast mode is outside the covered frequency range (Figure 4.2), τ_2 was fixed at 1.5 ps, the geometric average of the corresponding relaxation times of the pure ILs (1.22 ps for [emim][BF$_4$] and 1.84 ps [emim][DCA]).[38] The fit parameters and the corresponding χ_r^2 obtained from fitting the experimental spectra are summarized in Table 4.2.

4.2 Results

The DR spectra of neat imidazolium ILs are rather complicated. The $\hat{\varepsilon}(\nu)$ spectra are dominated by a relaxation centered at $\sim$3 GHz, which can be mainly attributed to the large-angle jump reorientation of the dipolar cations (Section 3.1). This is the major contribution to the present $\hat{\varepsilon}(\nu)$ in the region $0.2 \lesssim \nu/\text{GHz} \leq 20$ and is the focus of the following discussion.
Additionally, all imidazolium ILs studied so far show significant contributions from intermolecular vibrations and librations in the 0.1-10 THz region (Section 3.1). For ionic liquids having anions with sufficiently high dipole moments a further mode at $\sim$10 GHz can be detected.[37] The dicyanamide anion has a small dipole moment of $\mu \approx 1$ D,[49] however, as found in previous investigations covering larger frequency ranges,[38,39] no separate anion relaxation could be

4.2. RESULTS

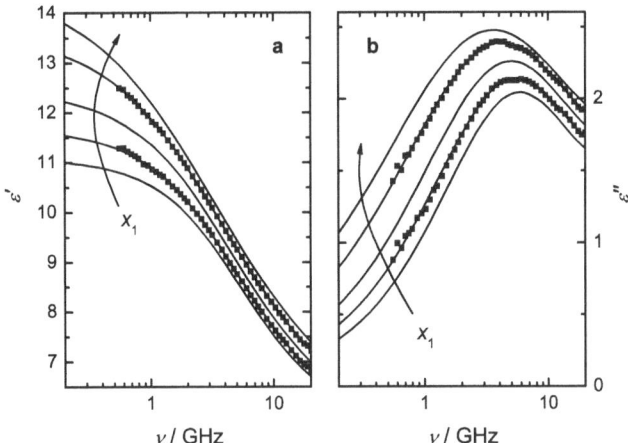

Figure 4.1: Spectra of (**a**) the dielectric permittivity, $\varepsilon'(\nu)$, and (**b**) the dielectric loss, $\varepsilon''(\nu)$, of selected [emim][BF$_4$] (**1**) + [emim][DCA] (**2**) mixtures at 25 °C. Symbols show representative experimental data, lines represent the CC + D fit. Arrows indicate increasing [emim][BF$_4$] content ($x_1 = 0$, 0.2164, 0.4740, 0.7856, 1).

detected for [emim][DCA]. On the other hand, as indicated by the superiority of the CC + D fit over the HN model in terms of χ_r^2, the low-frequency wing of the intermolecular vibrations significantly contributes to $\hat\varepsilon(\nu)$ (Figure 4.2) and was accordingly taken into account in the fitting procedure.

Compared with the $\hat\varepsilon(\nu)$ data for the neat ILs, covering the frequency range $0.2 \lesssim \nu/\text{GHz} \leq 89$,[38] the present spectra, limited to $\nu \leq 20\,\text{GHz}$, yield a slightly larger static permittivity. However, the increase was well within the estimated uncertainty: $0.6 \leq |\delta\varepsilon| \leq 1.5$ (Figure 4.3). Note that $\delta\varepsilon$ increases with decreasing [emim][DCA] mole fraction, x_2, due to the shift of the CC loss peak to lower frequencies (Figure 4.1). The corresponding variations of τ_1 and α_1 were also well within their estimated uncertainties. It can thus be safely concluded that the data of Table 4.2 yield reliable quantitative information on the cation relaxation of [emim][BF$_4$] + [emim][DCA] mixtures, and that extension of the frequency range to 89 GHz is not required.

The obtained static permittivities, ε (Figure 4.3), and relaxation times, τ_1, of the mixtures varied smoothly with concentration and were well described by second order polynomials

$$\varepsilon = 14.7 - 4.71\,x_2 + 1.10\,x_2^2, \qquad \sigma = 0.05 \tag{4.1}$$

Table 4.2: Fit Parameters of the CC + D Model, Eq. 1.61, for the DR Spectra at $0.2 \leq \nu/\text{GHz} \leq 20$ of [emim][BF$_4$] (**1**) + [emim][DCA] (**2**) Mixtures at 25 °C: Static Permittivity, ε; Relaxation Time, τ_1, Cole-Cole Width Parameter, α_1, Amplitudes, S_1 & S_2, Infinite-frequency Permittivity, ε_∞, and Reduced Error Function, χ_r^2. For all Fits the Relaxation Time of the High-frequency Debye Mode was Fixed to $\tau_2 = 1.5$ ps.

x_2	ε	τ_1/ps	α_1	S_1	S_2	ε_∞	$\chi_r^2/10^{-4}$
0	14.7	48.6	0.37	8.99	1.70	4.02	1.1
0.05527	14.5	46.4	0.36	8.81	1.63	4.05	1.2
0.2144	13.7	41.8	0.33	7.95	1.76	4.02	1.7
0.3746	13.1	37.6	0.29	7.25	1.81	4.02	1.8
0.5260	12.5	35.6	0.26	6.64	1.91	3.98	2.5
0.6782	12.1	33.8	0.24	6.14	1.97	3.94	2.7
0.7836	11.7	32.4	0.22	5.83	1.98	3.92	3.4
0.9552	11.2	30.6	0.19	5.30	2.11	3.83	4.0
1	11.1	30.4	0.18	5.15	2.07	3.89	3.9

and

$$\ln(\tau_1/\text{ps}) = 3.879 - 0.737\, x_2 + 0.276\, x_2^2, \qquad \sigma = 0.3 \qquad (4.2)$$

where x_2 is the [emim][DCA] mole fraction and σ is the standard deviation of the fit. For the relaxation-time distribution (broadness) parameter of the CC mode, α_1 (Table 4.2), a linear decrease from the value of neat [emim][BF$_4$] ($\alpha_1 = 0.37$) to that of neat [emim][DCA] ($\alpha_1 = 0.18$) was observed.

The measured densities, ρ, conductivities, κ, and corrected viscosities, η_corr, of the mixtures also followed second-order polynomials:

$$\rho/\text{g L}^{-1} = 1280.31 - 190.702\, x_2 + 12.3037\, x_2^2, \qquad \sigma = 0.06\,\text{g L}^{-1} \qquad (4.3)$$

$$\kappa/\text{S m}^{-1} = 1.552 + 1.584\, x_2 - 0.3031\, x_2^2, \qquad \sigma = 0.002\,\text{S m}^{-1} \qquad (4.4)$$

$$\eta_\text{corr}/10^{-3}\,\text{Pa s} = 36.8 - 29.3\, x_2 + 9.56\, x_2^2, \qquad \sigma = 0.6 \cdot 10^{-3}\,\text{Pa s} \qquad (4.5)$$

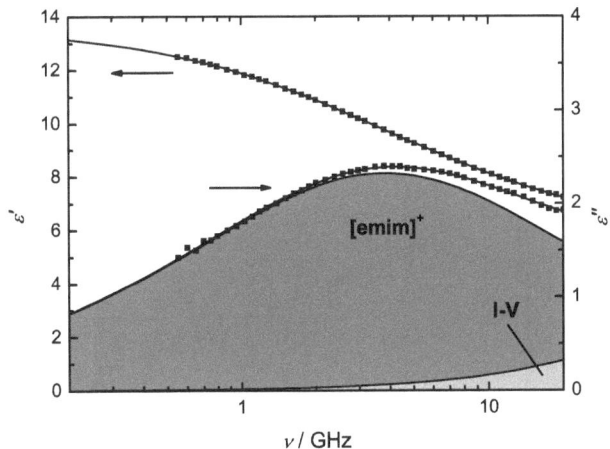

Figure 4.2: Dielectric permittivity, $\varepsilon'(\nu)$, and loss, $\varepsilon''(\nu)$, spectra of a [emim][BF$_4$] (**1**) + [emim][DCA] (**2**) mixture with $x_2 = 0.2144$ at 25 °C. Symbols show experimental data, lines represent the CC + D fit. The shaded areas indicate the contributions of the CC mode associated with cation reorientation ([emim]$^+$) and of the D mode arising from intermolecular vibrations (I-V).

4.3 Discussion

Excess properties

To infer specific intermolecular interactions between the components of binary mixtures it is common to use excess quantities, Y^E, which describe deviations from ideal mixing behavior. Strictly speaking, ideal (and hence excess) quantities are unambiguously defined only for thermodynamic properties. However, it is convenient to extend the concept to other properties, where 'ideal' behavior is usually taken to be 'simple linear mixing' with respect to a given concentration scale. For thermodynamic state properties the approach

$$Y^\text{E} = Y - Y^\text{id} = Y - \left(\sum_{i=1}^{2} x_i Y_i\right) \tag{4.6}$$

is appropriate, where Y and Y_i are the values of the thermodynamic property of the binary mixture and of the neat components, i, respectively, and the ideal property, Y^id, is defined by the bracketed term in Eq. 4.6. This leads for example to the relationship

$$V^\text{E} = \sum_{i=1}^{2} x_i M_i \left(\rho^{-1} - \rho_i^{-1}\right) \tag{4.7}$$

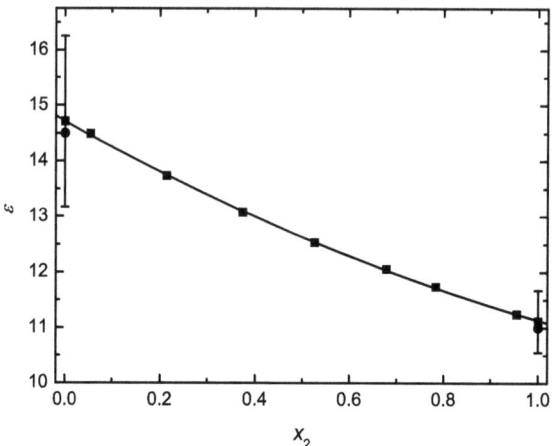

Figure 4.3: Static permittivity, ε, of [emim][BF$_4$] (**1**) + [emim][DCA] (**2**) mixtures at 25 °C (■) and data from ref. 38 (●) with estimated uncertainties for the neat ILs; the line represents a fit with Eq. 4.1.

between the excess molar volume, V^E, and the experimental densities of the mixture, ρ, and the pure components, ρ_i, where M_i is the molar mass of components i.

Excess permittivities, ε^E, excess conductivities, κ^E, and excess molar volumes, V^E, were calculated via Eqs. 4.6 and 4.7, respectively, from the corresponding data in Tables 4.1 and 4.2. Their relative values, Y^E/Y^{id}, are shown in Figure 4.4. Note that calculation of ε^E with Eq. 4.6 requires constant $\varepsilon_\infty(x_i)$.[170] This condition is sufficiently fulfilled here, see Table 4.2.

The viscosity of an ideal mixture is not uniquely defined and various empirical models can be found in the literature.[172,173] Within the accuracy of the present data (Figure 4.5), η shows 'ideal' behavior according to the definitions of Bingham,[174]

$$\frac{1}{\eta^{id}} = \sum_{i=1}^{2} \frac{x_i}{\eta_{i,\text{corr}}} \quad (4.8)$$

and of Grunberg and Nissan,[175]

$$\ln \eta^{id} = \sum_{i=1}^{2} x_i \ln \eta_{i,\text{corr}} \quad (4.9)$$

4.3. DISCUSSION

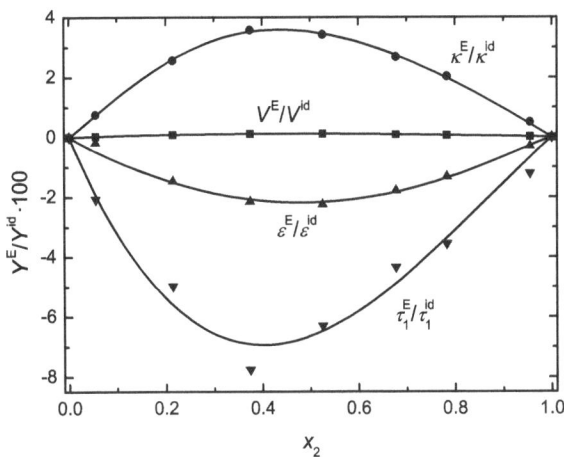

Figure 4.4: Relative excess properties, Y^E/Y^{id}, for molar volume, V^E/V^{id} (■), conductivity, κ^E/κ^{id} (●), relaxation time of the CC relaxation, τ_1^E/τ_1^{id} (▼), and static permittivity, $\varepsilon^E/\varepsilon^{id}$ (▲), of [emim][BF$_4$] (**1**) + [emim][DCA] (**2**) mixtures at 25 °C; lines are fits with a Padé approximation[171] and a guide to the eye only.

Neglecting the small effect of converting dielectric relaxation times to rotational correlation times, τ_1 should be proportional to viscosity if rotational diffusion of individual dipoles is the mechanism for the observed IL relaxation.[60] Thus, 'ideal' mixing behavior of τ_1 should follow the equation

$$\ln \tau_1^{id} = \sum_{i=1}^{2} x_i \ln \tau_{1,i} \qquad (4.10)$$

The corresponding relative excess relaxation times, $\tau_1^E/\tau_1^{id} = (\tau_1 - \tau_1^{id})/\tau_1^{id}$, are also included in Figure 4.4. Absolute values of τ_1^E obtained for the alternative definition of 'ideal' mixing,

$$(\tau_1^{id})^{-1} = \sum_{i=1}^{2} x_i \tau_{1,i}^{-1} \qquad (4.11)$$

were ∼40% smaller but nevertheless beyond experimental uncertainty.

As stated above, the present [emim][BF$_4$] (**1**) + [emim][DCA] (**2**) mixtures exhibited 'ideal' (simple linear) mixing for the width parameter, α_1, of the cation relaxation and for viscosity, η (Eqs. 4.8 or 4.9). The excess volume was always positive but values were very small ($V^E/V^{id} \leq 0.1\%$, Figure 4.4). Thus, the slight decrease of dipole density associated with V^E cannot explain the marked negative values of ε^E. Pronounced 'excess' properties peaking at $x_2 \approx 0.45$ were

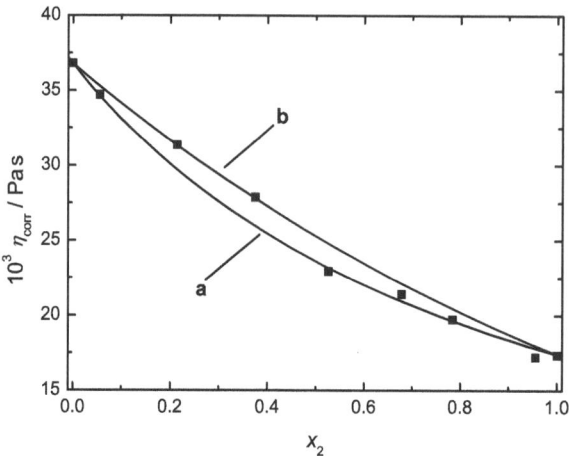

Figure 4.5: Corrected viscosities, η_{corr} (symbols), of [emim][BF$_4$] (**1**) + [emim][DCA] (**2**) mixtures together with 'ideal' viscosities (lines) calculated with Eqs. 4.8 (line **a**) and 4.9 (line **b**).

observed for conductivity, κ, and relaxation time, τ_1, with a maximum for the former and a minimum for the latter (Figure 4.4), suggesting enhanced translational (κ) and reorientational (τ_1) dynamics in the binary mixtures relative to the 'ideal' mixture behavior.

Interpretation of relaxation parameters

The investigated IL mixtures share a common cation ([emim]$^+$). The only differences among these mixtures—with possible implications on their structure and dynamics—must therefore arise via the anions. Tetrafluoroborate (BF$_4^-$) is spherical, with a van der Waals volume of $V_{\text{vdW}} = 47.7\,\text{Å}^3$ (MOPAC AM1 calculation),[176] whereas dicyanamide (DCA$^-$) is boomerang-shaped with $V_{\text{vdW}} = 55.7\,\text{Å}^3$. The surface-charge density of both ions is rather small and similar but, in contrast to BF$_4^-$, DCA$^-$ has a small permanent dipole moment of $\mu \approx 1\,\text{D}$ along its symmetry axis.[49] Thus, in addition to packing effects and Coulomb interactions among the ions, dipole-dipole interactions between DCA$^-$ and [emim]$^+$ ($\mu \approx 2\,\text{D}$)[49] are possible.

The spectra of the present mixtures and their pure components are dominated by the relaxation centered at $\sim 3-8\,\text{GHz}$ with parameters (S_1, τ_1, α_1). This mode can be unequivocally assigned to the cation.[36,38,48,177,178] However, even for the neat imidazolium ILs this mode has been shown not to be due to the uncorrelated rotational diffusion of individual cations but to reflect

4.3. DISCUSSION

static dipole-dipole correlations (*i.e.* structure) and cooperative dynamics, dominated by cation reorientation through large-angle jumps (Section 3.1). It should be noted that, whilst interactions in ILs and their mixtures are undoubtedly dominated by direct cation-anion contacts, the existence of individual long-lived ion pairs can be excluded.[46,47]

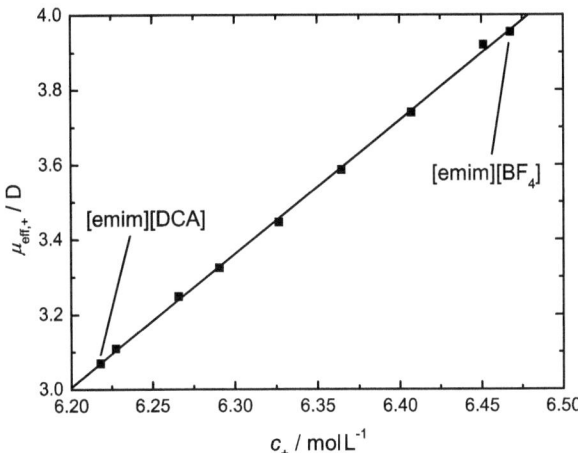

Figure 4.6: Effective dipole moments, $\mu_{\text{eff},+}$, of the CC relaxation of [emim][BF$_4$] + [emim][DCA] mixtures at 25 °C plotted as function of the molar concentration of cations, c_+, together with a straight-line fit of $\mu_{\text{eff},+} = f(c_+)$.

Some information on these interactions is accessible from the effective dipole moment, $\mu_{\text{eff},+}$, calculated from the amplitude S_1 of the observed relaxation via Eq. 1.68, where $c_j = c_+$, is the molar concentration of cations. Figure 4.6 shows that $\mu_{\text{eff},+}$ of the investigated mixtures increases linearly from 3.07 D for [emim][DCA] to 3.95 D for [emim][BF$_4$]. The dependence of S_1 on $\mu^2_{\text{eff},+}$, Eq. 1.68, explains why the quadratic expression, Eq. 4.1, was found for $\varepsilon(x_2)$ and why the resulting ε^{E} values are < 0 despite the negligible excess volumes (Figure 4.4). The linear change of $\mu_{\text{eff},+}$ implies a *smooth* transition from the liquid-state structure of [emim][DCA] to that of [emim][BF$_4$] and rules out structural features, such as well-defined aggregates, that would be *specific* to the binary mixtures but not to the pure ILs.
For imidazolium ILs containing a common cation, $\mu_{\text{eff},+}$ depends on the anion even when

the latter has zero dipole moment,[38] suggesting that the average relative orientation of the cations depends on the nature of the neighboring anions. Recent investigations of mixtures of [emim][EtSO$_4$] with DCM yielded $\mu_{\text{app},+} = 4.64\,\text{D}$ for [emim]$^+$ at infinite dilution and revealed strong parallel alignment ($g_+ = 2.7$, Eq. 1.71) of the cation dipoles and anti-parallel orientational correlations among the anions in the pure IL.[37] Using the above value for $\mu_{\text{app},+}$, correlation factors $g_+ = \mu_{\text{eff},+}^2/\mu_{\text{app},+}^2 = 0.44$ for [emim][DCA] and $g_+ = 0.72$ for [emim][BF$_4$] were obtained. Such values should not be over-interpreted, especially since the experimental g_+ factors almost certainly include cross-correlation effects among the different dipolar species (here [emim]$^+$ and DCA$^-$). Additionally, the values of $\mu_{\text{app},i}$ may vary somewhat with changing composition due to the change of the embedding environment. Nevertheless, these data suggest, for both ILs, a preference for anti-parallel alignment of [emim]$^+$ dipoles. The lack of a separate dicyanamide relaxation indicates that the dipole moments of DCA$^-$ essentially self-cancel in the liquid. This may either come from anti-parallel anion-anion alignment or, more likely, through cation-anion cross correlations.

Structural information inferred from dielectric data is inevitably indirect but it should be consistent with molecular-level scenarios provided by computer simulations. Recently, de Andrade et al.[179] performed a detailed analysis of the structure of liquid [emim][BF$_4$]. According to these simulations the local environment of the cations is characterized by the stacking of [emim]$^+$ in the z-direction perpendicular to the imidazolium rings, whereas the counterions are located in the equatorial planes. A marked preference for parallel and anti-parallel mutual orientations of the rings was observed, signifying pronounced stacking over several layers. Such a structure is compatible with the present dielectric data. According to the simulations, the dipole moments of the stacked cations show essentially random orientation in the xy-plane which contrasts somewhat with the present results as the correlation factor of $g = 0.72$ suggests a small preference for an anti-parallel alignment of cation moments. However, it should not be forgotten that the stacking-induced restriction of 'allowed' configurations of neighboring dipoles to essentially two dimensions affects g_+.

The salt [emim][DCA] was studied by Schröder et al.[180,181] These simulations yielded almost isotropic radial distribution functions and the authors concluded that no preferred mutual orientation exists in this IL. This contrasts markedly to the present result, where the value of $g_+ = 0.44$ hints at a rather pronounced mutual cancellation of dipole moments that almost certainly involves cations *and* anions. Although a quantitative comparison of experimental and simulated dielectric spectra is not appropriate, it should be noted that the simulations yielded a contribution from dipole reorientation, $S_1 \approx 3$, which is considerably smaller than the present experimental value of ~ 9 (Table 4.2). It is thus likely that the potential used by Schröder et

4.3. DISCUSSION

al.[180,181] does not adequately reproduce the structure of liquid [emim][DCA].
As indicated above, the linear variation of $\mu_{\text{eff},+}$ with mole fraction x_2 suggests a smooth transition of the IL structure going from the BF_4^- to the DCA^- salt. This also implies that the *local* structures of the neat liquids, sensed as dipole-dipole correlations by DR spectroscopy, should not be too different. Support for this conclusion comes from the linear ('ideal') variation of the Cole-Cole parameter, α_1. The dynamics of imidazolium ILs are thought to be heterogeneous and dominated by large-angle jumps of the dipolar cations, with α_1 being a measure for the spread of environments experienced by the relaxing dipoles.[38] According to the decreasing values of α_1 with increasing x_2 (see Table 4.2), [emim][DCA] is less heterogeneous than [emim][BF$_4$]. This is consistent with the inference from the g_+ values that the former should be relatively more structured compared to the latter, where g_+ is closer to unity.

However, this raises the question as to why η and τ_1 are smaller and κ larger for [emim][DCA]! That is, why translational and rotational dynamics are slower in [emim][BF$_4$]! A possible explanation comes from the simulations of de Andrade *et al.* for [emim][AlCl$_4$] which suggest that for this IL, stacked cation dimers with anions above and below are found instead of the large stacks dominating the structure of [emim][BF$_4$].[179] A similar dimeric structure for the dicyanamide would probably facilitate viscous flow, ion diffusion and cation reorientation and simultaneously allow anti-parallel dipole-dipole correlations in cation-cation *and* cation-anion pairs. One may speculate further that, in the mixture, the packing mismatch between large stacks and dimers of cations—possibly forming a 'random co-network' in the sense of Xiao *et al.*[9,10]—further increases the dynamics and leads to the observed excess effects for τ_1 and κ (Figure 4.4).

Chapter 5

IL + polar solvent mixtures

Parts of the material presented in this chapter were published in:

Stoppa, A.; Hunger, J.; Buchner, R. *'Conductivities of Binary Mixtures of Ionic Liquids with Polar Solvents.'* J. Chem. Eng. Data **2009**, *54*, 472.

and

Stoppa, A.; Hunger, J.; Hefter, G.; Buchner, R. *'Speciation in 1-Alkyl-3-Methylimidazolium Tetrafluoroborate + Acetonitrile Mixtures.'* in preparation.

5.1 Supplementary measurements

Precise and accurate values of the conductivities of all IL + solvent mixtures were determined using the setup described in Section 2.3. Additionally, the densities were measured, but for reasons that will be discussed, the values presented were not corrected for the damping effect of viscous samples.[119] They were used to calculate molar concentrations, c, molar conductivities, $\Lambda = \kappa/c$, and excess molar volumes, $V^{\rm E}$. Note, that all mixtures were studied over the whole composition range, except for [bmim][Cl] + AN mixtures ($0 < x_{\rm IL} \leq 0.07585$), as this IL is solid at room-temperature.

Tables 5.1 and 5.2 list the compositions of all investigated mixtures together with the values of κ, ρ, Λ and $V^{\rm E}$. Figure 5.1 shows the composition dependence of κ for [bmim][BF$_4$] + AN mixtures (which are representative for all the mixtures studied). For these mixtures the data presented here were those obtained using the IL [emim][BF$_4$]#1, as these were the mixtures studied by DR spectroscopy. The physico-chemical properties of [emim][BF$_4$]#2 + AN mixtures can be found in the literature.[22] Measurements on [bmim][BF$_4$] + AN mixtures were performed using the high-quality batch [bmim][BF$_4$]#2.

The values obtained here for κ, ρ and λ overcome, at least partly, the unsatisfactory lack of physico-chemical property data. To clarify, one may check the conductivity data for binary mixtures listed in the Ionic Liquids Database (ILThermo):[182] our ref. 22 is cited eleven times out of a total number of only 43 citations (last database update was 05/25/2010).

Table 5.1: Investigated Mole Fractions, x_{IL}, Corresponding Molar Concentrations, c_{IL} and c_{AN}, Densities, ρ, Conductivities, κ, Molar Conductivities, Λ, and Excess Molar Volumes, V^E, of Binary IL + AN Mixtures at 25 °C.

x_{IL}	c_{IL} $\overline{\text{mol L}^{-1}}$	c_{AN} $\overline{\text{mol L}^{-1}}$	ρ $\overline{\text{g L}^{-1}}$	κ $\overline{\text{S m}^{-1}}$	Λ $\overline{\text{S cm}^2\,\text{mol}^{-1}}$	$10^6\, V^E$ $\overline{\text{m}^3\,\text{mol}^{-1}}$
			[emim][BF$_4$]			
0.01082	0.2014	18.41	795.48	1.90	94.4	-0.212
0.01646	0.3036	18.14	804.68	2.55	84.0	-0.299
0.02252	0.4114	17.85	814.30	3.13	76.1	-0.385
0.04929	0.8621	16.63	853.38	4.89	56.7	-0.684
0.08059	1.342	15.30	893.87	6.03	44.9	-0.951
0.1218	1.903	13.72	939.88	6.72	35.3	-1.19
0.1736	2.513	11.96	988.57	6.92	27.6	-1.37
0.2371	3.146	10.12	1038.16	6.65	21.2	-1.51
0.3264	3.867	7.982	1093.25	5.90	15.26	-1.54
0.4565	4.672	5.562	1153.25	4.64	9.94	-1.42
0.6520	5.528	2.951	1215.55	3.10	5.61	-1.01
0.8336	6.092	1.216	1255.87	2.11	3.46	-0.506
1	6.484	0	1283.70	1.553	2.40	0

Conductivity. The variation of the electrical conductivity with IL content follows the typical pattern of concentrated electrolyte solutions: after a rapid initial rise at low x_{IL}, κ passes through a pronounced maximum at $x_{IL} \approx 0.1 - 0.25$ (depending on the IL). The appearance of such a curve is not surprising, given the two main effects that influence κ: the number of charge carriers and the viscosity of the mixtures, which determines the mobility of the charge carriers. Both effects scale with the IL content, but they have a counterbalancing effect onto

5.1. SUPPLEMENTARY MEASUREMENTS

Table 5.1 Continued.

x_{IL}	c_{IL} $\overline{\mathrm{mol\,L^{-1}}}$	c_{AN} $\overline{\mathrm{mol\,L^{-1}}}$	ρ $\overline{\mathrm{g\,L^{-1}}}$	κ $\overline{\mathrm{S\,m^{-1}}}$	Λ $\overline{\mathrm{S\,cm^2\,mol^{-1}}}$	$10^6\,V^{\mathrm{E}}$ $\overline{\mathrm{m^3\,mol^{-1}}}$
		[bmim][BF$_4$]				
0.003131	0.05883	18.73	782.18	0.698	118.6	-0.0550
0.006352	0.1185	18.54	787.76	1.215	102.5	-0.111
0.009414	0.1744	18.35	792.87	1.629	93.4	-0.157
0.01472	0.2696	18.05	801.77	2.22	82.4	-0.248
0.01981	0.3588	17.75	809.82	2.71	75.4	-0.316
0.02574	0.4601	17.42	818.95	3.14	68.1	-0.394
0.03142	0.5545	17.10	827.13	3.51	63.3	-0.444
0.03774	0.6570	16.75	836.23	3.84	58.5	-0.518
0.04389	0.7538	16.42	844.54	4.12	54.7	-0.567
0.05668	0.9472	15.77	861.33	4.55	48.0	-0.684
0.07179	1.162	15.03	879.61	4.88	42.0	-0.792
0.08907	1.392	14.23	898.88	5.10	36.6	-0.896
0.1077	1.622	13.43	918.10	5.19	32.0	-0.999
0.1538	2.123	11.68	959.09	5.04	23.8	-1.18
0.2149	2.666	9.740	1002.40	4.50	16.89	-1.29
0.2966	3.239	7.680	1047.32	3.62	11.16	-1.36
0.3900	3.741	5.853	1085.90	2.72	7.27	-1.33
0.6126	4.549	2.876	1146.22	1.285	2.82	-0.977
0.7544	4.893	1.593	1171.38	0.802	1.639	-0.644
0.8943	5.157	0.6095	1190.63	0.500	0.970	-0.310
1	5.319	0	1202.19	0.353	0.664	0
		[hmim][BF$_4$]				
0.004117	0.07700	18.63	784.24	0.838	108.8	-0.0865
0.008445	0.1560	18.32	791.63	1.442	92.5	-0.152
0.01744	0.3142	17.70	806.26	2.35	74.7	-0.276
0.03816	0.6489	16.36	836.42	3.51	54.2	-0.498
0.06486	1.028	14.82	869.76	4.13	40.2	-0.719
0.09763	1.427	13.19	903.86	4.25	29.8	-0.906
0.1399	1.855	11.41	939.67	3.96	21.4	-1.07
0.1943	2.300	9.539	975.94	3.37	14.67	-1.18
0.2712	2.785	7.484	1014.86	2.53	9.08	-1.25
0.3907	3.322	5.181	1056.74	1.543	4.65	-1.19
0.5902	3.894	2.704	1100.45	0.656	1.685	-0.917
0.7435	4.183	1.443	1122.12	0.344	0.823	-0.616
0.8889	4.387	0.5486	1137.20	0.192	0.437	-0.279
1	4.512	0	1146.34	0.1228	0.272	0

Table 5.1 Continued.

x_{IL}	c_{IL} $\overline{\mathrm{mol\,L^{-1}}}$	c_{AN} $\overline{\mathrm{mol\,L^{-1}}}$	ρ $\overline{\mathrm{g\,L^{-1}}}$	κ $\overline{\mathrm{S\,m^{-1}}}$	Λ $\overline{\mathrm{S\,cm^2\,mol^{-1}}}$	$10^6\,V^{E}$ $\overline{\mathrm{m^3\,mol^{-1}}}$
			[bmim][PF$_6$]			
0.007571	0.1404	18.41	795.63	1.485	105.8	-0.119
0.01576	0.2863	17.88	815.17	2.54	88.7	-0.236
0.02489	0.4415	17.30	835.56	3.39	76.8	-0.337
0.03448	0.5970	16.72	855.89	4.05	67.9	-0.436
0.05838	0.9529	15.37	901.77	5.00	52.5	-0.632
0.08779	1.338	13.90	950.71	5.39	40.3	-0.818
0.1261	1.766	12.24	1004.47	5.27	29.8	-0.988
0.1772	2.238	10.39	1062.70	4.68	20.9	-1.12
0.2521	2.780	8.246	1128.61	3.63	13.04	-1.20
0.3649	3.374	5.871	1199.71	2.30	6.82	-1.19
0.5591	4.038	3.184	1278.14	0.995	2.47	-0.952
0.7291	4.415	1.640	1321.97	0.469	1.062	-0.598
0.8729	4.652	0.6772	1349.76	0.253	0.543	-0.359
1	4.815	0	1368.32	0.1469	0.305	0
			[bmim][Cl]			
0.002633	0.04959	18.79	779.97	0.382	77.0	-0.0594
0.005508	0.1033	18.64	783.44	0.611	59.1	-0.117
0.01261	0.2335	18.29	791.60	0.984	42.1	-0.241
0.02572	0.4658	17.64	805.63	1.403	30.1	-0.434
0.03837	0.6799	17.04	818.22	1.652	24.3	-0.595
0.05100	0.8843	16.46	829.99	1.818	20.6	-0.738
0.07585	1.261	15.36	850.93	1.98	15.71	-0.960
			[bmim][DCA]			
0.01992	0.3602	17.73	801.64	2.86	79.4	-0.374
0.04752	0.8063	16.16	828.91	4.62	57.3	-0.614
0.07852	1.245	14.61	855.13	5.54	44.5	-0.837
0.1175	1.720	12.91	883.10	5.95	34.6	-1.08
0.1647	2.201	11.16	909.96	5.88	26.7	-1.23
0.2294	2.737	9.192	939.06	5.37	19.6	-1.35
0.3205	3.315	7.028	969.04	4.50	13.58	-1.35
0.4353	3.854	4.999	996.22	3.49	9.06	-1.24
0.6384	4.498	2.548	1027.88	2.25	4.99	-0.900
0.7724	4.794	1.413	1042.05	1.716	3.58	-0.597
0.8946	5.008	0.5899	1052.20	1.369	2.73	-0.304
1	5.160	0	1059.21	1.139	2.21	0

5.1. SUPPLEMENTARY MEASUREMENTS

Table 5.1 Continued.

x_{IL}	c_{IL} $\overline{\text{mol L}^{-1}}$	c_{AN} $\overline{\text{mol L}^{-1}}$	ρ $\overline{\text{g L}^{-1}}$	κ $\overline{\text{S m}^{-1}}$	Λ $\overline{\text{S cm}^2 \text{mol}^{-1}}$	$10^6 \, V^{\text{E}}$ $\overline{\text{m}^3 \text{mol}^{-1}}$
			[hmim][NTf$_2$]			
0.002346	0.04386	18.65	785.45	0.490	111.7	-0.0174
0.004808	0.08896	18.41	795.68	0.891	100.1	-0.120
0.007385	0.1347	18.11	803.63	1.247	92.5	-0.0552
0.01009	0.1818	17.84	813.84	1.573	86.5	-0.131
0.01592	0.2788	17.24	832.32	2.13	76.5	-0.108
0.02244	0.3819	16.64	853.83	2.62	68.7	-0.223
0.03795	0.6036	15.30	898.25	3.34	55.3	-0.348
0.05770	0.8463	13.82	946.04	3.73	44.1	-0.433
0.08403	1.116	12.17	999.12	3.80	34.1	-0.534
0.1229	1.433	10.23	1060.81	3.51	24.5	-0.638
0.1776	1.765	8.172	1125.19	2.92	16.53	-0.719
0.2595	2.111	6.025	1192.00	2.10	9.93	-0.802
0.3417	2.351	4.528	1237.60	1.506	6.41	-0.781
0.4541	2.578	3.099	1280.51	0.985	3.82	-0.674
0.5969	2.773	1.873	1317.40	0.613	2.21	-0.548
0.8171	2.964	0.6635	1353.32	0.330	1.115	-0.262
1	3.069	0	1372.92	0.218	0.709	0
			[emim][EtSO$_4$]			
0.008943	0.1659	18.38	793.84	1.023	61.7	-0.177
0.01904	0.3452	17.79	811.71	1.602	46.4	-0.328
0.04008	0.6934	16.61	845.63	2.31	33.3	-0.584
0.06873	1.118	15.14	885.68	2.81	25.1	-0.833
0.1036	1.568	13.56	927.28	3.03	19.35	-1.06
0.1482	2.057	11.82	971.36	3.04	14.77	-1.25
0.2100	2.613	9.828	1020.98	2.79	10.68	-1.45
0.2861	3.151	7.862	1067.33	2.36	7.50	-1.51
0.3866	3.692	5.858	1112.79	1.822	4.94	-1.45
0.6063	4.480	2.909	1177.94	0.999	2.23	-1.12
0.7645	4.851	1.494	1207.54	0.666	1.374	-0.679
0.8924	5.082	0.6129	1225.89	0.494	0.973	-0.307
1	5.242	0	1238.69	0.392	0.748	0

Table 5.2: Investigated Mole Fractions, x_{IL}, Corresponding Molar Concentrations, c_{IL} and c_{MeOH}, Densities, ρ, Conductivities, κ, and Molar Conductivities, Λ, of Binary [bmim][BF$_4$] + MeOH Mixtures at 25 °C.

x_{IL}	c_{IL} $\overline{\text{mol L}^{-1}}$	c_{MeOH} $\overline{\text{mol L}^{-1}}$	ρ $\overline{\text{g L}^{-1}}$	κ $\overline{\text{S m}^{-1}}$	Λ $\overline{\text{S cm}^2\text{mol}^{-1}}$	$10^6\, V^{\text{E}}$ $\overline{\text{m}^3\text{mol}^{-1}}$
0.007410	0.1776	23.79	802.48	0.867	48.8	-0.102
0.01543	0.3603	23.00	818.28	1.451	40.3	-0.188
0.02435	0.5526	22.14	834.29	1.97	35.6	-0.250
0.03355	0.7396	21.30	849.76	2.40	32.4	-0.307
0.05717	1.173	19.35	885.02	3.30	28.1	-0.415
0.08589	1.624	17.29	921.13	3.81	23.4	-0.507
0.1242	2.127	14.99	961.01	4.18	19.65	-0.609
0.1755	2.662	12.51	1002.48	4.26	15.99	-0.655
0.2480	3.237	9.817	1046.31	3.92	12.11	-0.654
0.3624	3.877	6.820	1094.85	3.09	7.98	-0.629
0.5583	4.556	3.604	1145.28	1.760	3.863	-0.417
1	5.319	0	1202.19	0.353	0.664	0

5.1. SUPPLEMENTARY MEASUREMENTS

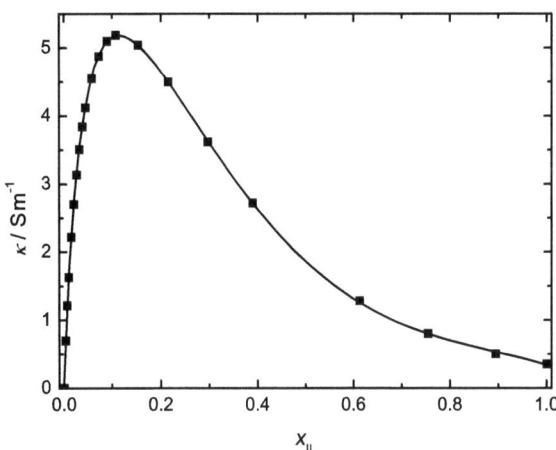

Figure 5.1: Conductivities, κ, of [bmim][BF$_4$] + AN mixtures at 25 °C. The line shows a fit of Eq. 5.1 to the present data.

conductivity. Though these effects are well understood, a theoretical prediction of the transport properties of such systems is still not possible. However, a recent approach based on lattice-hole theory was applied to explain κ of IL + solvent mixtures, in particular water mixtures, with acceptable accuracy.[183]

To reproduce the composition dependence of the experimental κ values an empirical Padé (n, m) equation,

$$Y = x_{\text{IL}}(1 - x_{\text{IL}}) \frac{\sum_{i=1}^{n} A_i (2x_{\text{IL}} - 1)^{i-1}}{1 + \sum_{j=1}^{m} B_j (2x_{\text{IL}} - 1)^j} + C_1 x_{\text{IL}} \tag{5.1}$$

can be used, where $Y = \kappa$. In Eq. 5.1, $\kappa = 0$ is assumed for the pure solvent (valid within our error limits for κ), whereas parameter C_1 represents the conductivity of the pure IL. Best results were obtained by using a seven parameter fit with $n = 4$ and $m = 2$ (except for [bmim][Cl] + AN, where $n = 2$, $m = 2$ and $C_1 = 0$ produced an excellent fit). The parameters A_i, B_j and C_1 obtained by least-squares analysis are summarized in Table 5.3, together with the corresponding standard deviations of the fits, σ. The line in Figure 5.1 represents such a fit. In the fitting procedure C_1 was adjusted but always found to be close to the measured data (compare Tables 5.1 and 5.3).

Molar conductivity. For the investigated mixtures, it was found that in general Λ decreases continuously with increasing IL content (Tables 5.1 and 5.2). Probably, this reflects decreasing ion mobility induced by rising viscosity.

Table 5.3: Parameters A_i, B_j and C_1, Together with the Corresponding Standard Deviations, σ, of Fits of Eq. 5.1 (with $Y = \kappa$, $n = 4$ and $m = 2$) to the Conductivity, κ, as a Function of IL Mole Fraction, x_{IL}, of IL + Solvent Mixtures at 25 °C.

	A_1	A_2	A_3	A_4	B_1	B_2	C_1	$10^2\,\sigma$
	$\overline{\text{S m}^{-1}}$	$\overline{\text{S m}^{-1}}$	$\overline{\text{S m}^{-1}}$	$\overline{\text{S m}^{-1}}$			$\overline{\text{S m}^{-1}}$	$\overline{\text{S m}^{-1}}$
[emim][BF$_4$] + AN	13.93	4.199	-5.645	3.190	1.786	0.7897	1.550	0.8
[bmim][BF$_4$] + AN	6.786	-1.078	-2.091	5.510	1.870	0.8708	0.3369	2
[hmim][BF$_4$] + AN	3.527	-7.840	9.645	-5.574	0.2445	-0.6255	0.1233	4
[bmim][PF$_6$] + AN	4.769	-9.404	11.02	-6.483	0.4603	-0.3931	0.1478	4
[bmim][Cl] + AN	5.605	5.556	0	0	1.935	0.9352	0	0.6
[bmim][DCA] + AN	9.763	2.706	-3.298	4.669	1.926	0.9211	1.137	2
[hmim][NTf$_2$] + AN	2.899	-2.862	3.962	-3.913	1.031	0.09827	0.2186	2
[emim][EtSO$_4$] + AN	4.536	0.01109	-1.413	2.833	1.845	0.8460	0.3853	1
[bmim][BF$_4$] + MeOH	7.763	0.3209	-5.068	2.238	1.810	0.8107	0.3534	3

The data were fit to the empirical equation

$$\Lambda = D_1 - D_2\sqrt{c_{\text{IL}}} + D_3 c_{\text{IL}} \ln c_{\text{IL}} + D_4 c_{\text{IL}}, \tag{5.2}$$

which corresponds to the truncated series expansion derived for dilute electrolyte solutions from the theory of Onsager,[184,185] but without a theoretical meaning for the parameters D_i, $i = 1\ldots 4$. The parameters obtained and the corresponding σ values are summarized in Table 5.4.

It was previously shown,[22] that the agreement among the limited data for κ and Λ of IL + solvent mixtures is poor. The present results for the pure ILs conform excellently with *some* of the published data. Electrode polarization and sample purity were discussed in detail in ref. 22 as possible sources of error. To quantify the latter, measurements of AN mixtures prepared with either [emim][BF$_4$]#1 (commercial sample) or [emim][BF$_4$]#2 (home-made sample), were compared. They showed systematic deviations of up to 3.4 % with an average difference of 1.3 %,[22] which exceeded the experimental accuracy of ca. 0.5 %. The impact of BF$_4^-$ hydrolysis, which seems to have been overlooked by some authors, was discussed. It was suggested, that

5.1. SUPPLEMENTARY MEASUREMENTS

Table 5.4: Parameters D_i of Fits of Eq. 5.2 to Molar Conductivities, Λ, as a Function of Molar IL Concentration, c_{IL}, at 25 °C. The Standard Deviations, σ, of the Fits are Indicated.

	$\dfrac{10^4\,D_1}{\mathrm{S\,m^2\,mol^{-1}}}$	$\dfrac{10^4\,D_2}{\mathrm{S\,m^{7/2}\,mol^{-3/2}}}$	$\dfrac{10^6\,D_3}{\mathrm{S\,m^5\,mol^{-2}}}$	$\dfrac{10^5\,D_4}{\mathrm{S\,m^5\,mol^{-2}}}$	$\dfrac{10^5\,\sigma}{\mathrm{S\,m^2\,mol^{-1}}}$
[emim][BF$_4$] + AN	148.2	5.089	-1.340	1.582	6
[bmim][BF$_4$] + AN	156.0	6.202	-1.802	2.102	10
[hmim][BF$_4$] + AN	149.5	5.649	-1.235	1.548	8
[bmim][PF$_6$] + AN	146.9	3.794	0.04067	0.2067	5
[bmim][Cl] + AN	140.8	13.55	-10.72	10.48	7
[bmim][DCA] + AN	137.8	3.853	-0.4932	0.6947	3
[hmim][NTf$_2$] + AN	134.8	3.790	0.1829	0.09734	8
[emim][EtSO$_4$] + AN	107.0	4.983	-1.742	1.976	7
[bmim][BF$_4$] + MeOH	74.29	2.810	-1.151	1.233	7

some of the data found in the literature are flawed by major contaminations of the studied samples. Much more experimental care has to be taken and the purity of the ILs should be properly determined and declared.[186]

Density. The situation is also unsatisfactory regarding the densities of neat ILs and IL + solvent mixtures. The available literature data for [bmim][BF$_4$] + AN mixtures[187–191] are compared with the present (uncorrected) values in Figure 5.2. Agreement between the present results and those of Zafarani-Moattar and Shekaari[188,191] over the whole composition range and those of Wang et al.[187] at $x_{IL} \leq 0.6$ are quite satisfactory with $|\delta(\rho)| = |\rho_{\mathrm{present}} - \rho_{\mathrm{literature}}| \leq 2\,\mathrm{g\,L^{-1}}$. However, there are marked disagreements in the ρ values reported by other workers, especially at higher x_{IL}, with $|\delta(\rho)|$ increasing up to $20\,\mathrm{g\,L^{-1}}$ (Figure 5.2). Unfortunately, this situation is common for ILs and usually reflects the presence of impurities, especially water or halides.[186]

Another factor influencing the accuracy of measurements with vibrating-tube density meters is the damping effect of viscous samples, i.e. the oscillation period of the vibrating tube depends on sample viscosity, η, leading to systematic errors in ρ.[119] As discussed in ref. 22, this effect is negligible for low-viscosity samples, but pure ILs and IL + IL mixtures are generally quite viscous, and the systematic effect exceeds the stated accuracy. Applying the procedure described by Heintz et al.[171] we checked the magnitude of the viscosity correction for [bmim][BF$_4$] + AN

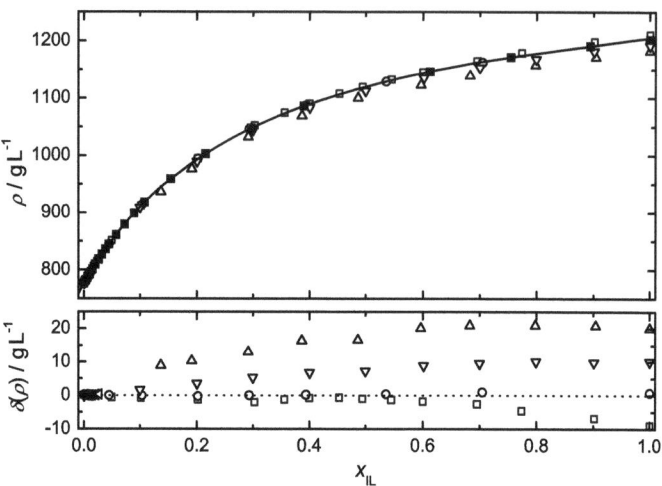

Figure 5.2: Experimental densities, ρ, of mixtures of [bmim][BF$_4$] and acetonitrile at 25 °C together with absolute deviations, $\delta(\rho) = \rho_{\text{present}} - \rho_{\text{literature}}$, as a function of the IL mole fraction, x_{IL}: ■, this work; □, ref. 187; ○, ref. 188; △, ref. 189; ▽, ref. 190; ◁, ref. 191. Lines are included as visual guides.

mixtures,[22] various neat ILs[97] and IL + IL mixtures.[192] For $\eta \gtrsim 6\,\text{mPa\,s}$ the effect of η on ρ is larger than the claimed uncertainty of the measurement ($\pm 0.05\,\text{kg\,m}^{-3}$). The magnitude of the effect is well below the uncertainty of some related quantities (like molar conductivities or in the analysis of DR spectra), whereas others are significantly affected: relative deviations of uncorrected, V^{E}, and corrected excess molar volumes, $V^{\text{E}}_{\text{corr}}$, were estimated for [emim][BF$_4$] + [emim][DCA] mixtures using measured viscosities and the procedure described by Heintz et al.[171] yielding $(V^{\text{E}} - V^{\text{E}}_{\text{corr}})/V^{\text{E}}_{\text{corr}} \lesssim 4\,\%$.

State-of-the-art equipment uses a reference oscillation tube to automatically correct for the damping effect, as realized in the DMA 5000 M density meter (Section 2.3.1). This instrument was used for the measurement of ρ for selected [emim][EtSO$_4$] + AN mixtures (Figure 5.3 and Appendix A.1). The agreement between values determined with the DMA 5000 M and interpolated values measured with the DMA 60, DMA 601HT density meter is quite satisfactory

5.1. SUPPLEMENTARY MEASUREMENTS

(with $|\delta(\rho)| = |\rho_{\text{DMA60}} - \rho_{\text{DMA5000M}}| \leq 1.3\,\text{g}\,\text{L}^{-1}$). These measurements can give an impression of the magnitude of the damping effect, indicating that discrepancies of $|\delta(\rho)| \approx 20\,\text{g}\,\text{L}^{-1}$ (see above) must be due to the presence of impurities, either inherently present in the samples or introduced by inadequate sample handling.

As data for the viscosities of selected [emim][EtSO$_4$] + AN mixtures are available (see Appendix A.1), the procedure for the correction of the damping effect described by Heintz et al.[171] could be checked. Figure 5.3 shows that generally $|\delta(\rho)| = |\rho_{\text{DMA60,corrected}} - \rho_{\text{DMA5000M}}|$ decreases somewhat after use of the correction procedure, thus justifying its application. Note, that the value of $\delta(\rho) = -1.1\,\text{g}\,\text{L}^{-1}$ at $x_{\text{IL}} = 0.04004$ cannot be explained on the basis of viscous damping, as $\delta(\rho) > 0$ would be expected.[171] Possibly, and purely speculatively, evaporation of AN during the filling procedure lead to an increase in η.

Nevertheless, due to the lack of reliable viscosity data for the majority of mixtures, and in the absence of more reliable and sophisticated correction procedures, uncorrected values of ρ are given in this thesis.

Excess molar volumes, V^{E}, at 25 °C were calculated via

$$V^{\text{E}} = x_{\text{IL}} M_{\text{IL}} (\rho^{-1} - \rho_{\text{IL}}^{-1}) + (1 - x_{\text{IL}}) M_{\text{solvent}} (\rho^{-1} - \rho_{\text{solvent}}^{-1}) \qquad (5.3)$$

All values of ρ and ρ_{IL} were taken from the present measurements, except for the IL [bmim][Cl] ($\rho_{\text{[bmim][Cl]}} = 1080\,\text{g}\,\text{L}^{-1}$),[193] and the solvents AN ($\rho_{\text{AN}} = 776.669\,\text{g}\,\text{L}^{-1}$)[54] and MeOH ($\rho_{\text{MeOH}} = 786.373\,\text{g}\,\text{L}^{-1}$).[57] The values so derived are listed in Tables 5.1 & 5.2; typical data are shown in Figure 5.4 for selected IL + AN systems. Due to the damping effect of viscous samples, the values derived are possibly only accurate to $\sim$4 % (estimated for IL + IL mixtures). Nevertheless, the qualitative discussion of V^{E} will not be affected. The V^{E} values for all of the studied IL + solvent mixtures show negative deviations from ideal ($V^{\text{E}} \equiv 0$) mixing and a pronounced asymmetry of the $V^{\text{E}}(x_{\text{IL}})$ curves.

To reproduce $V^{\text{E}}(x_{\text{IL}})$, fits using Padé equations (Eq. 5.1 with $Y = V^{\text{E}}$) were performed. In Eq. 5.1, $C_1 = 0$ throughout and hence $V^{\text{E}} = 0$ is obtained for the pure components. In the fitting procedure the best results were obtained by using a five parameter fit with $n = 2$ and $m = 3$ (except for [bmim][Cl] + AN, where $n = 2$ and $m = 2$ gave an excellent fit). The parameters A_i and B_j obtained by least-squares analysis are summarized in Table 5.5, together with the corresponding standard deviations of the fits, σ. The lines in Figure 5.4 represent such a fit.

The asymmetry of the curves is consistent with the tendency of ILs to maintain their IL-like 'character' up to relatively high levels of dilution in IL + solvent mixtures (Section 5.2).

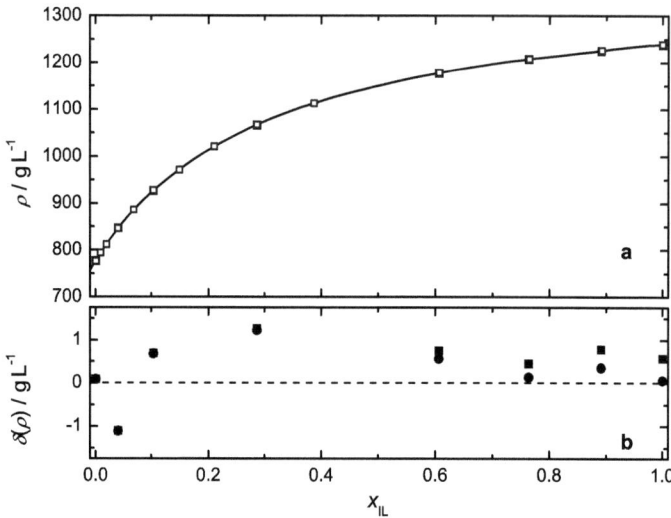

Figure 5.3: (a) Experimental densities, ρ, of mixtures of [emim][EtSO$_4$] and AN at 25 °C measured with a DMA 60, DMA 601HT (open symbols) and a DMA 5000 M density meter (full symbols, hardly distinguishable) as a function of the IL mole fraction, x_{IL}. The line is a polynomial fit to the DMA 60 data, which was used for interpolation. (b) Absolute deviations, $\delta(\rho)$ (see text) of the DMA 5000 M measurements from uncorrected (■) and corrected (●) values obtained with the DMA 60 density meter.

Another interesting feature of the data is that the minima of $V^E(x_{IL})$ show systematic behavior with respect to the solvent: for all IL + AN mixtures the minima are at $x_{IL} = 0.29 \pm 0.02$, being independent of the choice of cation and/or anion, but for [bmim][BF$_4$] + MeOH the minimum is shifted to $x_{IL} \approx 0.23$. Moreover, the data for IL + AN mixtures show a systematic trend with increasing chain-length of the 1-hydrocarbon substituent on the imidazolium ring (Figure 5.4). In general, V^E([hmim][X]) > V^E([bmim][X]) > V^E([emim][X]) over the entire composition range, i.e., the variation of the IL anion (X$^-$) has only minor effects on V^E (NTf$_2^-$ being the only exception). Independent measurements will be needed to establish the validity (or otherwise) of these effects. Nevertheless, for the present mixtures the results suggest that the properties of IL + AN mixtures show little dependence on the choice of the IL.

5.2. DIELECTRIC PROPERTIES

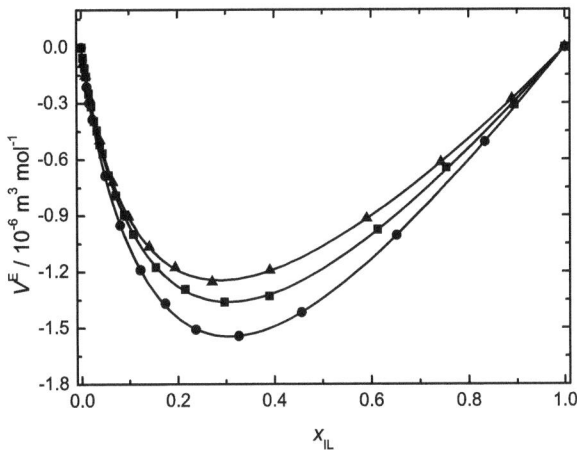

Figure 5.4: Excess molar volumes, V^E, of binary IL + AN mixtures at 25 °C as a function of the IL mole fraction, x_{IL}: ●, [emim][BF$_4$]; ■, [bmim][BF$_4$]; ▲, [hmim][BF$_4$]. Lines are fits of Eq. 5.1 to the present data.

5.2 Dielectric properties

This section deals with the analysis and interpretation of the DR spectra of various binary IL + polar solvent mixtures (see Introduction for an overview). It is subdivided into two parts: first, mixtures of AN and ILs containing symmetric, nonpolar (Cl$^-$, BF$_4^-$ and PF$_6^-$) or dipolar (DCA$^-$, NTf$_2^-$ and EtSO$_4^-$) anions will be discussed. The differences of [emim][EtSO$_4$] + AN mixtures compared to the others will be highlighted. The second part will deal with [bmim][BF$_4$] + MeOH mixtures.

5.2.1 IL + acetonitrile mixtures

Figure 5.5 shows the rather broad and featureless DR spectra of representative [bmim][BF$_4$] + AN mixtures, including the pure components.[38] The corresponding spectra for all IL + AN mixtures were rather similar. Because of this high degree of similarity the following discussion will focus on [bmim][BF$_4$] + AN mixtures as representative of the data as a whole. The

Table 5.5: Parameters A_i and B_j of Fits of Eq. 5.1 (with $Y = V^E$, $n = 2$, $m = 3$ and $C_1 = 0$) to Excess Molar Volumes, V^E, as a Function of IL Mole Fraction, x_{IL}, of IL + Solvent Mixtures at 25 °C. The Standard Deviations, σ, of the Fits are Indicated.

	$10^6 A_1$	$10^6 A_2$	B_1	B_2	B_3	$10^8 \sigma$
	$\mathrm{m^3\,mol^{-1}}$	$\mathrm{m^3\,mol^{-1}}$				$\mathrm{m^3\,mol^{-1}}$
[emim][BF$_4$] + AN	-5.369	-4.794	1.577	0.6407	0.03848	0.5
[bmim][BF$_4$] + AN	-4.742	-4.078	1.527	0.5722	0.01124	0.8
[hmim][BF$_4$] + AN	-4.260	-4.197	1.669	0.6255	-0.04574	0.4
[bmim][PF$_6$] + AN	-4.157	-4.104	1.663	0.6160	-0.04992	2
[bmim][Cl] + AN	-5.349	-5.264	1.697	0.6999	0	0.1
[bmim][DCA] + AN	-4.656	-4.844	1.728	0.6764	-0.04054	1
[hmim][NTf$_2$] + AN	-2.606	-2.821	1.805	0.7448	-0.04035	3
[emim][EtSO$_4$] + AN	-5.275	-4.649	1.568	0.6578	0.06207	2
[bmim][BF$_4$] + MeOH	-1.933	-1.746	1.968	1.262	0.2835	1

equivalent data for the other systems is provided in the following Tables and, where necessary, discrepancies will be highlighted in the text and visualized in comparative figures.

Fit model

The dielectric spectra for all IL + AN mixtures showed a more-or-less smooth transition from neat AN to the neat IL (representative data are shown for [bmim][BF$_4$] + AN in Figure 5.5). Nevertheless, the retrieval of the 'true' relaxation mechanism in these mixtures is not trivial. An extensive analysis using Eq. 1.61 indicated that several alternative superpositions of individual relaxation modes described the experimental spectra almost equally well, namely the CC + D, CC + D + D, D + CC + D and D + D + D models. Among the possibilities, a CC + D + D model fit the spectra of mixtures of [emim][EtSO$_4$] with AN best. For [emim][BF$_4$], [bmim][BF$_4$], [hmim][BF$_4$], [bmim][PF$_6$], [bmim][Cl], [bmim][DCA] and [hmim][NTf$_2$] + AN mixtures, a superposition of only two processes, a CC + D model, described the spectra very well. As shown in Section 1.3.5, consideration of the χ_r^2 parameter alone is not sufficient to rule out inappropriate models. However, the CC + D + D and CC + D models yielded the lowest values of χ_r^2 for the majority of spectra in the specified systems and the parameters obtained (Tables 5.6 and 5.7) were physically reasonable and varied smoothly with respect to composition.

5.2. DIELECTRIC PROPERTIES

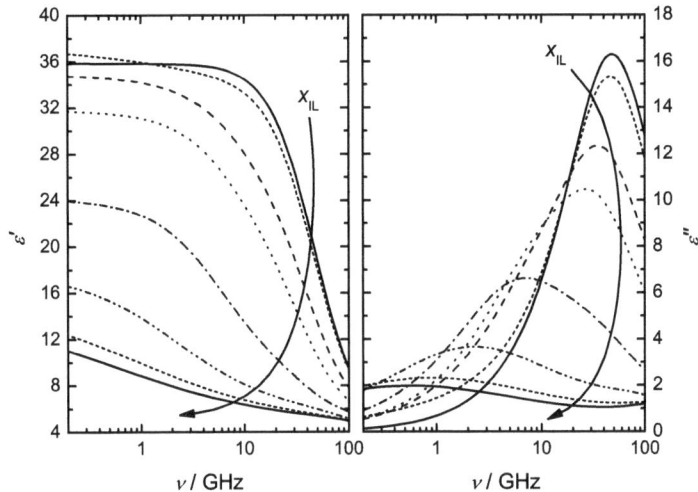

Figure 5.5: Dielectric permittivity, $\varepsilon'(\nu)$, and loss, $\varepsilon''(\nu)$, spectra of selected [bmim][BF$_4$] + AN mixtures at 25 °C. Arrows indicate increasing IL content (x_{IL} = 0, 0.006352, 0.05688, 0.1077, 0.2966, 0.6126, 0.8943, 1). Spectra of the neat components are indicated by full lines.

The same models were also adopted previously for [bmim][BF$_4$] + DCM and [emim][EtSO$_4$] + DCM mixtures.[37,48] Figure 5.6 shows the spectra of two representative [bmim][BF$_4$] + AN mixtures, along with their constituent CC and D contributions. Further analysis showed, that for spectra at low x_{IL} an additional small-amplitude Debye relaxation centered at ∼2 GHz (the D + CC + D model, see below) could be resolved. Due to strong mode overlap, it was not possible to fit the corresponding D + CC + D + D model to spectra of [emim][EtSO$_4$] + AN mixtures.

Composition dependence of relaxation parameters

High-frequency mode. In the present IL + AN mixtures, the location of the higher-frequency Debye mode, here labelled process 2, depends on x_{IL} (compare Figures 5.6a and 5.6b), with its center varying from ∼30 to ∼260 GHz. The composition dependence of τ_2 and S_2 is shown for [bmim][BF$_4$] + AN mixtures in Figure 5.7. Almost identical results were ob-

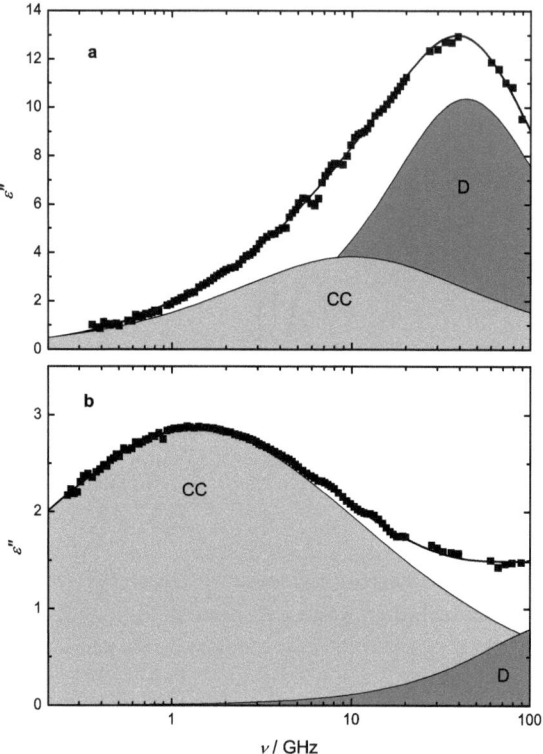

Figure 5.6: Dielectric loss spectra, $\varepsilon''(\nu)$, of two representative [bmim][BF$_4$] + AN mixtures at 25 °C with IL mole fractions $x_{IL} = 0.04389$ (**a**) and 0.7544 (**b**). Symbols represent experimental data, lines show the CC + D fit, and shaded areas indicate the contributions of the individual processes.

tained for the other systems. The relaxation times (Figure 5.7b) increase from the value of neat AN, $\tau_{AN} = 3.32$ ps, to a maximum of ~5 ps at $x_{IL} \approx 0.2$, with a subsequent smooth decrease until τ_2 approaches the value of the neat IL, $\tau_{IL} \approx 0.6$ ps obtained for a fit of the pure [bmim][BF$_4$] spectrum in the same frequency range.[38] The corresponding amplitudes, S_2, of the mixtures (Figure 5.7a) decrease with increasing IL content, pass through a poorly defined minimum at

5.2. DIELECTRIC PROPERTIES

$x_{IL} \approx 0.4$, and then approach practically linearly $S_{IL} = S_2(x_{IL} = 1)$. This behavior strongly indicates that process 2 is a superposition of modes arising from the ILs and the solvent. It is emphasized that, as for the IL + DCM mixtures studied previously,[37,48] the description of the dielectric spectra of the present IL + AN mixtures in terms of two or three apparent processes (i.e., the CC + D and CC + D + D models) masks a deeper complexity, as will be discussed in detail below.

Low-frequency mode. The variation of the average relaxation time of the low-frequency CC process, τ_1, and the corresponding broadness parameter, α, with composition were also essentially the same for all IL + AN systems studied. Representative values (for [bmim][BF$_4$] + AN) are presented in Figure 5.7, which shows that τ_1 decreases strongly upon small additions of the IL and exhibits a minimum at $x_{IL} \approx 0.08$ for [bmim][BF$_4$] + AN (Figure 5.7c). Similar-shaped curves were obtained for the CC broadness parameter but with the minimum at $x_{IL} \approx 0.2$ (Figure 5.7d). The amplitude of process 1, S_1, is shown as a function of x_{IL} for [bmim][BF$_4$] + AN mixtures in Figure 5.8. For all of the studied mixtures S_1 exhibited a strong increase with increasing x_{IL}, showing a shallow maximum at $x_{IL} \approx 0.2$ followed by a steady decline to the values found in the neat ILs. This is not so for [emim][EtSO$_4$] + AN mixtures, where S_1 exhibited a smooth increase with x_{IL} over the whole composition range (Figure 5.9a). The reasons for the obvious deviations will be discussed below. The concentration dependence of the static permittivities of the IL + AN mixtures was characterized by a weak maximum at $x_{IL} \approx 0.01$ and a subsequent decrease to the values of the neat ILs. For [emim][EtSO$_4$] +AN mixtures, a shallow minimum appears at $x_{IL} \approx 0.6$ (Figure 5.10).

Medium-frequency mode. In [emim][EtSO$_4$] + AN mixtures only, amongst the present systems, an additional mode, labelled here as the 'medium-frequency mode', is observed. The center of this medium-frequency mode, here identified as 'anion' process, varies from ~ 7 to ~ 18 GHz. After a smooth increase at low x_{IL}, τ_{anion} approaches the value of the neat IL without further variation with respect to IL content. Note, that for $x_{IL} \gtrsim 0.5$, τ_{anion} has practically the same values as the corresponding relaxation times in [emim][EtSO$_4$] + DCM mixtures,[37] but they are much less at lower x_{IL}. The behavior at high x_{IL} indicates, as found previously, that this process is due to the reorientation of the highly dipolar anion ($\mu_{ap,EtSO_4^-} = 11.2 - 13.2$ D, depending on its conformation).[37]

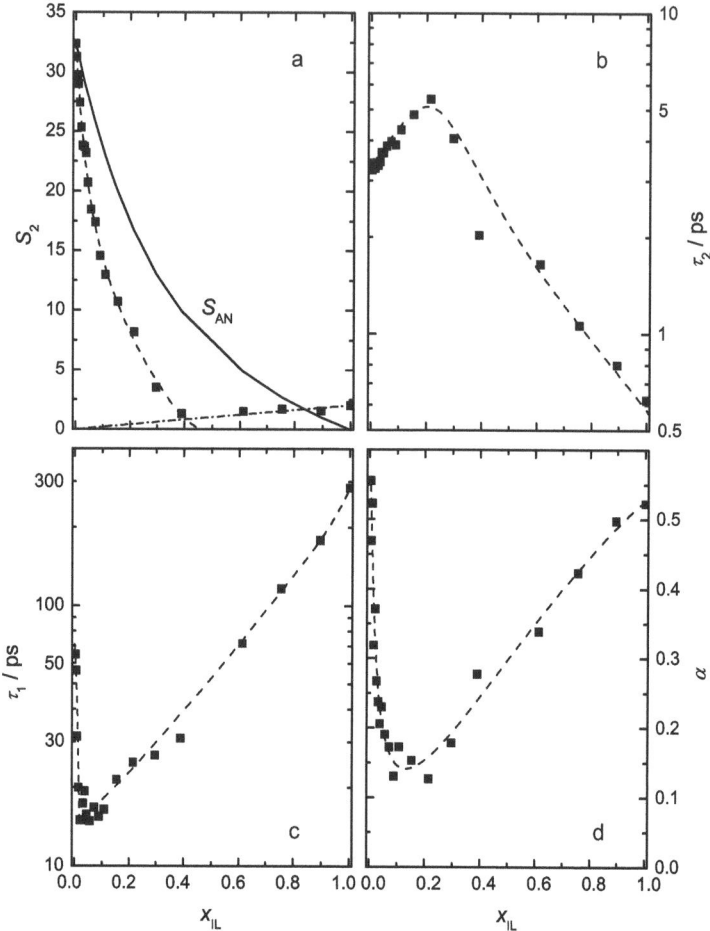

Figure 5.7: (a) Experimental amplitude of the high-frequency process, S_2 (symbols), and the AN relaxation estimated from the Cavell equation, S_{AN} (Eq. 1.68, full line), together with (b), the corresponding relaxation time, τ_2, and (c), the average relaxation time of process 1, τ_1, with (d), the corresponding Cole-Cole broadness parameter, α, for [bmim][BF$_4$] + AN mixtures at 25 °C as a function of IL mole fraction. Dashed lines are included only as a visual guide.

5.2. DIELECTRIC PROPERTIES

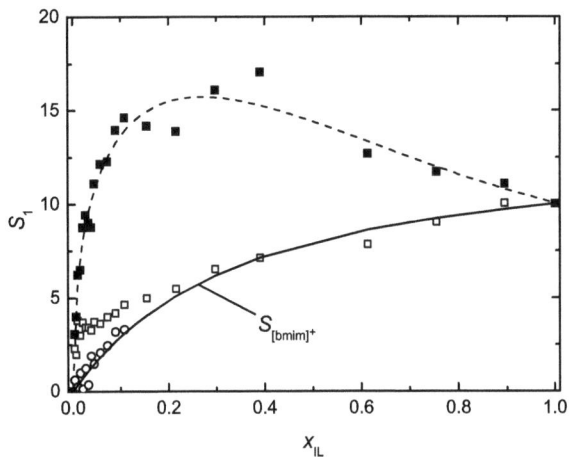

Figure 5.8: Amplitudes of relaxation process 1, S_1, (filled symbols, dotted line) and the cation relaxation, $S_{[bmim]^+}$, estimated from Eq. 1.68, (full line), of [bmim][BF$_4$] + AN mixtures at 25 °C together with reduced amplitudes, $S_{1,red}$, calculated for: $\square$, CC + D; and $\bigcirc$, D + CC + D models.

Discussion

High-frequency mode. The dielectric relaxation of neat AN is due to rotational diffusion of its molecular dipoles and therefore the SED theory is applicable (Section 3.2). According to this model, Eq. 1.77, the rotational correlation time, τ'_j, is a linear function of η/T for a given system. For neat AN, the values of V_m and f have been given in Section 3.2. Experimental macroscopic relaxation times, τ_j, can be converted to τ'_j via the Madden-Kivelson relation (Eq. 1.80). If τ'_j is known, Eq. 1.77 can be used 'in reverse' to calculate values of C and $V_{\text{eff},j}$ (Eq. 1.79). The latter can then be compared to quantum chemical calculations (see below).

The values of τ'_2 for [bmim][BF$_4$], [bmim][PF$_6$] and [emim][EtSO$_4$] + AN mixtures correlate well with solution viscosity (data obtained by interpolation of values given in refs. 187, 194 and

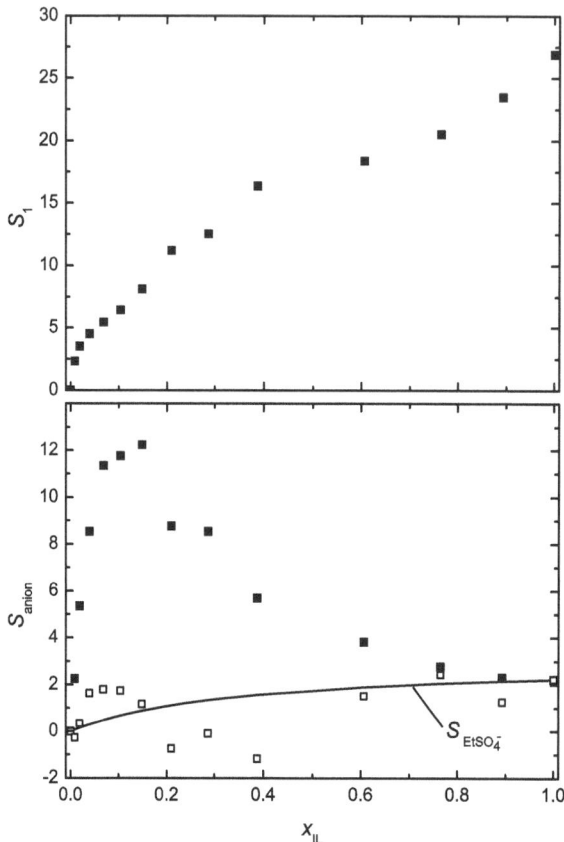

Figure 5.9: Experimental amplitudes of the low- and medium-frequency processes, S_1 and S_{anion}, respectively, together with the anion amplitude estimated from Eq. 1.68, $S_{\text{EtSO}_4^-}$ (full line), for [emim][EtSO$_4$] + AN mixtures at 25 °C assuming the CC + D + D model. Open symbols show reduced amplitudes, $S_{\text{anion,red}}$ (see text).

5.2. DIELECTRIC PROPERTIES

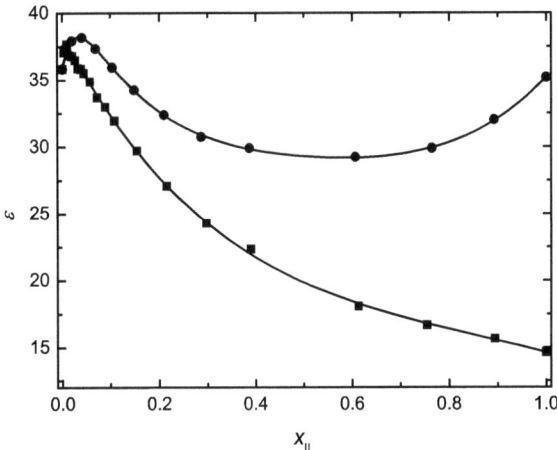

Figure 5.10: Static permittivities, ε, of binary IL + AN mixtures at 25 °C plotted as function of the IL mole fraction, x_{IL}, for the ILs: ■, [bmim][BF$_4$]; and ●, [emim][EtSO$_4$]. Lines are included as visual guides.

Table 5.6: Fit Parameters of Eq. 1.61 for the Observed Dielectric Spectra of Mixtures of ILs with AN at 25 °C Assuming the CC + D Model: Static Permittivities, ε; Relaxation Times, τ_j, and Amplitudes, S_j, of Process j; Cole-Cole Shape Parameter, α, of the First (Lower Frequency) Process; Infinite Frequency Permittivity, ε_∞, and Reduced Error Function of the Overall Fit, χ_r^2.

x_{IL}	ε	S_1	τ_1/ps	α	S_2	τ_2/ps	ε_∞	$\chi_r^2/10^{-4}$
0^a	35.84				32.5	3.32	3.33	
[emim][BF$_4$]								
0.01082	37.31	3.99	60.5	0.31	29.6	3.46	3.69	121
0.01646	37.17	5.18	35.0	0.29	28.2	3.49	3.79	182
0.02252	37.16	7.03	25^b	0.29	26.8	3.44	3.38	319
0.04929	36.18	11.5	15^b	0.23	21.1	3.62	3.57	201
0.08059	34.91	14.9	12.5	0.22	15.5	4.38	4.53	179
0.1218	32.79	17.2	12.7	0.16	11.4	4.08	4.23	142
0.1736	30.46	17.0	15^b	0.13	9.05	4.02	4.42	137
0.2371	28.15	17.1	17^b	0.15	6.12	4.95	4.89	130
0.3264	25.17	16.6	20.4	0.14	3.72	3.53	4.82	57.9
0.4565	21.97	14.5	27.8	0.16	2.80	2.58	4.65	28.2
0.6520	18.71	12.0	40.8	0.23	2.34	1.87	4.34	25.4
0.8336	16.63	10.5	50.5	0.31	1.96	1.66	4.20	6.20
1^c	14.5	8.70	46.6	0.36	2.05	1.22	3.75	

^{a} Parameters taken from Section 3.2.
^{b} Parameter fixed during fitting procedure.
^{c} Parameters from ref. 38.

Appendix A.1) at $x_{IL} \lesssim 0.15$ (Figure 5.11). The values of τ_2' are well fit by the linear equations

[bmim][BF$_4$] + AN: $\tau_2' = (1.81 \pm 0.05)\,\text{ps} + \eta \cdot (1288 \pm 75)\,\text{ps/Pa}\cdot\text{s}, \quad \sigma = 0.07\,\text{ps}$ (5.4)

[bmim][PF$_6$] + AN: $\tau_2' = (1.94 \pm 0.10)\,\text{ps} + \eta \cdot (1321 \pm 190)\,\text{ps/Pa}\cdot\text{s}, \quad \sigma = 0.1\,\text{ps}$ (5.5)

[emim][EtSO$_4$] + AN: $\tau_2' = (2.13 \pm 0.05)\,\text{ps} + \eta \cdot (452 \pm 51)\,\text{ps/Pa}\cdot\text{s}, \quad \sigma = 0.09\,\text{ps}$ (5.6)

Unfortunately, no data for the viscosities were available for the other systems studied. Application of Eq. 1.77 to the present data gives $C = 0.033$ for [bmim][BF$_4$] + AN, $C = 0.034$ for [bmim][PF$_6$] + AN and $C = 0.012$ for [emim][EtSO$_4$] + AN, which are somewhat smaller than

Table 5.6 Continued.

x_{IL}	ε	S_1	τ_1/ps	α	S_2	τ_2/ps	ε_∞	$\chi_r^2/10^{-4}$
			[bmim][BF$_4$]					
0.003131	37.09	3.08	64.7	0.56	31.3	3.25	2.74	133
0.006352	37.33	4.01	56.3	0.47	29.7	3.41	3.63	151
0.009414	37.69	6.22	31.4	0.52	29.0	3.35	2.47	144
0.01472	36.84	6.49	20^a	0.32	27.4	3.28	2.95	330
0.01981	36.79	8.75	15^a	0.37	25.3	3.41	2.72	231
0.02574	36.49	9.42	15^a	0.27	23.8	3.35	3.26	446
0.03142	35.88	8.99	17.4	0.24	23.7	3.44	3.21	245
0.03774	35.85	8.76	19.3	0.21	23.2	3.67	3.90	641
0.04389	35.53	11.1	15.7	0.23	20.7	3.66	3.71	289
0.05668	34.90	12.1	14.9	0.19	18.5	3.84	4.30	595
0.07179	33.70	12.3	16.7	0.17	17.4	3.97	4.02	186
0.08907	33.00	14.0	15.5	0.13	14.6	3.88	4.49	618
0.1077	31.95	14.6	16.5	0.17	13.0	4.33	4.36	167
0.1538	29.72	14.2	21.5	0.15	10.7	4.82	4.82	123
0.2149	27.08	13.9	25^c	0.13	8.17	5.39	5.02	140
0.2966	24.30	16.1	26.6	0.18	3.50	4.06	4.72	56.1
0.3900	22.35	17.1	30.9	0.28	1.30	2.03	3.99	36.3
0.6126	18.07	12.7	71.5	0.34	1.52	1.65	3.86	8.18
0.7544	16.67	11.7	116	0.42	1.71	1.06	3.24	9.32
0.8943	15.65	11.1	178	0.50	1.56	0.800	3.01	8.83
1^b	14.6	10.0	284	0.52	2.04	0.620	2.57	
			[hmim][BF$_4$]					
0.004117	36.91	2.47	116	0.33	30.8	3.36	3.64	174
0.008445	36.98	4.59	39.1	0.37	29.1	3.34	3.24	145
0.01744	36.66	5.53	37.6	0.24	27.2	3.59	3.95	221
0.03816	35.76	9.55	23.9	0.25	22.1	3.92	4.15	270
0.06486	34.34	14.8	16.9	0.25	16.3	3.76	3.21	157
0.09763	32.48	17.4	17.1	0.27	11.6	4.31	3.56	81.9
0.1399	30.55	20.0	17.1	0.28	6.88	5.08	3.70	88.9
0.1943	27.78	18.3	25^a	0.26	5.32	4.98	4.15	76.2
0.2712	25.28	17.3	35^a	0.30	3.62	7.23	4.39	80.1
0.3907	21.40	16.4	46.6	0.33	0.904	3.05	4.10	10.1
0.5902	17.24	12.5	109	0.40	1.09	1.93	3.69	15.4
0.7435	14.50	10.1	169	0.45	0.996	1.31	3.45	7.28
0.8889	12.84	8.72	262	0.52	1.35	0.576	2.77	6.34
1^b	12.0	7.87	451	0.54	1.95	0.44	2.18	

^{a} Parameter fixed during fitting procedure.

^{b} Parameters from ref. 38.

Table 5.6 Continued.

x_{IL}	ε	S_1	τ_1/ps	α	S_2	τ_2/ps	ε_∞	$\chi_r^2/10^{-4}$
\multicolumn{9}{c}{[bmim][PF$_6$]}
0.007570	36.60	3.36	60.0	0.35^a	30.3	3.28	2.97	225
0.01576	36.24	5.10	34.8	0.37	27.7	3.54	3.43	430
0.02489	35.09	5.81	24.1	0.2^a	25.9	3.48	3.43	501
0.03448	34.58	10.2	12.0	0.28	21.4	3.55	3.04	325
0.05838	33.07	15.0	9.91	0.26	15.1	3.78	3.02	309
0.08779	31.35	18.0	10.1	0.24	9.74	4.32	3.56	460
0.1261	29.34	17.2	14.0	0.21	8.20	4.33	3.94	265
0.1772	27.21	19.4	15.3	0.22	3.89	3.81	3.91	154
0.2521	24.80	20.2	18.9	0.26	0.578	3.5^a	4.02	152
0.3649	22.02	17.0	36.0	0.30	0.962	2.54	4.04	30.2
0.5591	19.73	15.2	111	0.42	1.12	1.29	3.38	16.7
0.7291	19.40	15.0	384	0.50	1.17	1.22	3.23	8.65
0.8729	17.19	12.8	679	0.53	0.987	1.42	3.39	12.9
1^b	16.1	12.0	1178	0.57	1.86	0.47	2.24	
\multicolumn{9}{c}{[bmim][Cl]}
0.002633	36.55	1.50	50^a	0.32	31.7	3.31	3.36	126
0.005508	37.14	2.69	46.7	0.23	31.0	3.37	3.45	389
0.01261	38.20	4.92	44.7	0.17	29.6	3.50	3.69	268
0.02572	39.63	8.97	39.9	0.21	26.8	3.60	3.88	255
0.03837	39.79	9.87	43.2	0.15	25.5	3.95	4.43	465
0.05100	40.27	13.8	35.6	0.24	22.6	3.86	3.95	497
0.07585	39.56	14.9	39.0	0.21	19.9	4.46	4.72	595
\multicolumn{9}{c}{[bmim][DCA]}
0.01992	37.10	9.56	14.1	0.41	25.1	3.47	2.40	265
0.04752	34.97	10.3	17.0	0.22	21.0	3.76	3.63	311
0.07852	33.24	18.5	9.54	0.26	11.6	4.23	3.12	128
0.1175	30.90	18.8	11.1	0.25	8.66	4.77	3.45	76.3
0.1647	28.40	17.4	14.2	0.24	6.88	5.93	4.15	70.7
0.2294	25.79	18.6	14.7	0.28	3.44	9.21	3.78	157
0.3205	22.46	18.3	18.3	0.26	0.205	1^a	3.91	47.7
0.4353	19.54	15.0	27.3	0.28	1.29	0.594	3.21	22.7
0.6384	15.68	10.7	48.1	0.29	1.41	1.23	3.58	9.01
0.7724	14.24	9.36	62.1	0.33	1.44	1.13	3.44	5.08
0.8946	12.95	8.13	73.8	0.35	1.43	1.09	3.40	3.15
1^b	11.31	6.55	61.7	0.34	1.23	1.22	3.53	

^{a} Parameter fixed during fitting procedure.

^{b} Parameters from ref. 38.

5.2. DIELECTRIC PROPERTIES

Table 5.6 Continued.

x_{IL}	ε	S_1	τ_1/ps	α	S_2	τ_2/ps	ε_∞	$\chi_r^2/10^{-4}$
				[hmim][NTf$_2$]				
0.002346	37.02	2.36	269	0.46	31.2	3.35	3.51	186
0.004810	36.61	2.32	278	0.18	30.9	3.40	3.44	1350
0.007385	36.30	3.48	65.0	0.3^a	29.4	3.38	3.39	335
0.01009	36.16	4.22	63.4	0.32	28.7	3.42	3.28	385
0.01592	35.24	4.62	46.4	0.2^a	26.8	3.55	3.82	285
0.02244	34.83	5.94	47.3	0.23	25.1	3.65	3.75	232
0.03795	33.04	8.34	32.5	0.25^a	20.8	3.91	3.88	432
0.05770	31.79	9.00	46.6	0.22	18.3	4.56	4.49	481
0.08403	30.48	12.8	40.5	0.3^a	13.5	4.85	4.12	366
0.1229	28.65	13.7	55.9	0.3^a	10.7	5.91	4.25	340
0.1776	26.38	14.5	73.6	0.3^a	7.71	7.02	4.19	269
0.2595	22.22	13.9	72.5	0.3^a	4.17	8.06	4.11	204
0.3417	20.08	14.5	81.7	0.35^a	1.89	7.15	3.66	153
0.4541	17.03	13.5	76.0	0.4^a	0.571	1.41	2.97	66.7
0.5969	14.81	11.4	108	0.43	0.938	0.740	2.52	34.5
0.8171	13.14	9.91	207	0.50	0.692	0.8^a	2.53	51.3
1^b	12.7	9.40	233	0.47	0.68	0.80	2.58	

^{a} Parameter fixed during fitting procedure.
^{b} Parameters from ref. 38.

Table 5.7: Fit Parameters of Eq. 1.61 for the Observed Dielectric Spectra of Mixtures of [emim][EtSO$_4$] with AN at 25 °C Assuming the CC + D + D Model: Static Permittivities, ε; Relaxation Times, τ_j, and Amplitudes, S_j, of Process j; Cole-Cole Shape Parameter, α, of the First (Lower Frequency) Process; Infinite Frequency Permittivity, ε_∞, and Reduced Error Function of the Overall Fit, χ_r^2.

x_{IL}	ε	S_1	τ_1/ps	α	S_{anion}	τ_{anion}/ps	S_2	τ_2/ps	ε_∞	$\chi_r^2/10^{-4}$
0.008940	37.23	2.32	69.9	0.07	2.25	9.09	28.9	3.36	3.72	205
0.01904	37.94	3.52	61.9	0.04	5.35	8.84	25.4	3.29	3.63	261
0.04008	38.19	4.51	58.2	0.03	8.53	10.6	21.5	3.24	3.61	235
0.06873	37.37	5.44	50.3	0.02	11.3	10.6	16.4	3.43	4.22	212
0.1036	35.95	6.42	44.6	0.02^a	11.8	11.7	13.2	3.71	4.59	217
0.1482	34.25	8.10	42.5	0.02	12.2	12.1	9.15	3.64	4.77	141
0.2100	32.39	11.2	43.6	0.04	8.77	14^a	7.28	4.44	5.15	99.0
0.2861	30.76	12.5	56.0	0.03	8.53	16^a	4.80	3.61	4.90	51.1
0.3866	29.92	16.3	74.2	0.09	5.68	17^a	3.15	2.88	4.74	36.0
0.6063	29.24	18.4	175	0.12	3.82	23.2	2.65	2.44	4.38	37.8
0.7645	29.90	20.5	299	0.17	2.77	23.8	2.20	2.49	4.40	24.1
0.8924	32.02	23.5	500	0.23	2.31	22.6	2.00	2.05	4.22	21.5
1^b	35.20	26.9	806	0.24	2.22	24.0	1.89	2.03	4.24	

^{a} Parameter fixed during fitting procedure.
^{b} Parameters from ref. 37.

the value of $C = 0.119$ for neat AN (Section 3.2) and the values of $C = 0.068$ and $C = 0.063$ given by Barthel *et al.* for Bu$_4$NBr and NaI solutions in AN.[85] This indicates that, compared to the pure solvent, rotational motions of AN molecules are weakly coupled to the shear stress in Bu$_4$NBr and NaI solutions, and even less coupled in [bmim][BF$_4$], [bmim][PF$_6$] and [emim][EtSO$_4$] + AN mixtures. This may be understood as follows: the linearity of $\tau_2' = f(\eta)$ suggests diffusive reorientational motions of the AN molecules. As will be shown below, interactions of AN with the cations are relatively weak and limited to the first solvation shell. Thus, the reorientational motions of a major portion of the AN molecules are not affected by the presence of ions and behave essentially like 'bulk' AN. MD simulations have indeed shown, that interactions between AN molecules are enhanced after mixing with [bmim][BF$_4$],[31] consequently leading to structural heterogeneity.[32,33] Nevertheless, one should keep in mind the

5.2. DIELECTRIC PROPERTIES 91

possible influence of orientational or dynamical correlations, that is $\dot{g}/g_K \neq 1$ in Eq. 1.81, and the superposition of AN and IL modes in process 2, which could reduce the observed relaxation time τ_2 to a considerable extent.

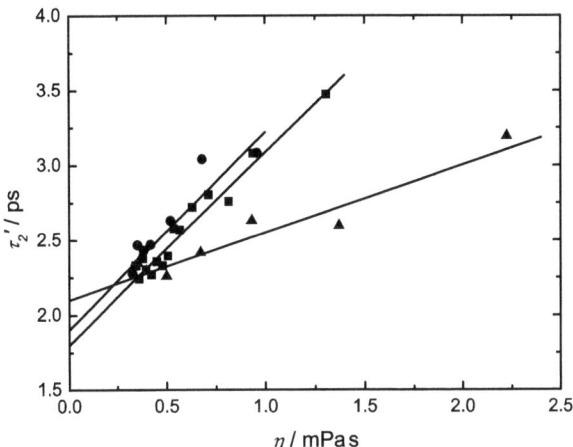

Figure 5.11: Microscopic relaxation time of process 2, τ_2', of binary IL + AN mixtures at 25 °C plotted as function of solution viscosity for the ILs: ■, [bmim][BF$_4$]; ●, [bmim][PF$_6$]; ▲, [emim][EtSO$_4$]. The lines are linear fits in the range $0 \leq x_{IL} \leq 0.1539$, ≤ 0.1261 and ≤ 0.2100, respectively.

Figure 5.7a compares the experimental values of S_2 with S_{AN}. The values of the latter were estimated from Eq. 1.68 assuming that all AN molecules contribute to the spectrum and that their dipole moment, $\mu_{\text{eff,AN}}$, has the same value as in neat AN. Two features of the amplitudes in Figure 5.7a are important. First, the experimental amplitudes are well below the expected ones, i.e., $S_2 < S_{AN}$, at $x_{IL} \lesssim 0.4$ indicating that with IL addition a decreasing portion of AN molecules contributes to process 2, or in other words, AN molecules are 'bound' (slowed down on the DS timescale) and thus no longer contribute to process 2. Second, S_2 increases slightly at $x_{IL} \gtrsim 0.4$ until it eventually becomes $> S_{AN}$. Since this is a physical impossibility it must mean, as observed for IL + DCM mixtures,[37,48] that one or more IL modes contribute to the observed amplitude S_2. This is strongly supported by the observation that the values of S_2 for

$x_{IL} \gtrsim 0.4$ are on a straight line connecting (0,0) and the point for the pure IL (within the scatter of the data true for all mixtures studied). The difference between S_{AN} and S_2 for $x_{IL} \lesssim 0.4$ not only yields reasonable solvation numbers but also allows a consistent interpretation of the low-frequency amplitude S_1 in terms of bound AN, free cations and contact ion pairs (see below).

Effective solvation numbers, Z_b, were calculated via

$$Z_b = \frac{c_{AN} - c_{AN,app}}{c_{IL}} \qquad (5.7)$$

where $c_{AN,app}$ is the apparent concentration of AN calculated from S_2 and Eq. 1.68 assuming that the effective dipole moment of AN in the mixtures is the same as in pure AN. The linearly increasing contribution of the IL mode was not taken into account, as the effects on Z_b are much below the scatter of the data. Typical results are plotted in Figure 5.12 for [emim][BF$_4$], [bmim][BF$_4$] and [hmim][BF$_4$] + AN mixtures. Due to the relatively modest ion-solvation properties of AN (with donor and acceptor numbers of 14.1 and 19.3, respectively),[195] cations should be weakly solvated by AN, whereas anions should be practically unsolvated.[196,197] The values of Z_b plotted as function of x_{IL} are virtually independent of the nature of the hydrocarbon substituent on the imidazolium ring, thus indicating that the polar part of the cation is decisive for solvation.

A comparison of all mixtures studied showed a high similarity in the variation of Z_b with respect to IL content and it was found, that $\ln Z_b$ varies linearly with IL concentration in the range $0 \leq c_{IL}/\text{mol L}^{-1} \lesssim 3$, i.e. fits according to

$$\ln Z_b = \ln Z_b^\circ - b \cdot c_{IL} \qquad (5.8)$$

were performed.

For all mixtures, the infinite-dilution solvation numbers, Z_b°, were $\sim 5 - 10$ (Table 5.8), which indicate the formation of a complete solvation shell. The scatter of Z_b° is in the range $\Delta Z_b^\circ \approx \pm 1$, but one has to keep in mind, that the error due to the extrapolation (limited concentration range, effect of overlapping modes) will further increase the overall uncertainty. Particularly at low x_{IL}, the scatter of Z_b is high, making a reliable extrapolation difficult. Thus, the values of Z_b° should not be overinterpreted, nevertheless, no general trend with respect to the cation is discernible. The high value of $Z_b^\circ = 9.7$ for [emim][EtSO$_4$] + AN mixtures may be a consequence of possible anion-AN interactions due to the large dipole moment of EtSO$_4^-$.

Comparison of the solvation number obtained for Bu$_4$NBr in AN, $Z_b^\circ = 2.7$,[198] with the present results suggests that imidazolium-based cations are more strongly solvated than Bu$_4$N$^+$.

At high x_{IL}, $S_2 > S_{AN}$. As explained above, this physically impossible result occurs because of overlapping AN and IL modes, with the latter eventually contributing the major part. Since

5.2. DIELECTRIC PROPERTIES

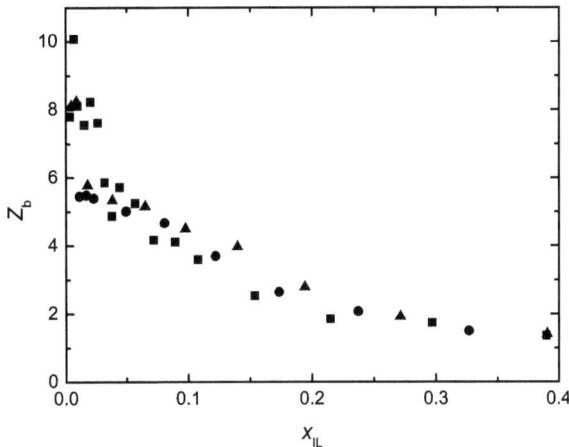

Figure 5.12: Solvation numbers, Z_b, of binary IL + AN mixtures at 25 °C as a function of the IL mole fraction, x_{IL}: ●, [emim][BF$_4$]; ■, [bmim][BF$_4$]; ▲, [hmim][BF$_4$].

Table 5.8: Infinite-Dilution Solvation Numbers, Z_b°, for IL + AN mixtures at 25 °C.

mixture	Z_b°
[emim][BF$_4$]	6.2
[bmim][BF$_4$]	9.0
[hmim][BF$_4$]	7.6
[bmim][PF$_6$]	6.7
[bmim][Cl]	5.3
[bmim][DCA]	9.1
[hmim][NTf$_2$]	7.1
[emim][EtSO$_4$]	9.7

in the ~100 GHz region at least two IL relaxations are present in the DR spectra,[39] a further analysis is not justified on the basis of spectra recorded at $\nu \leq 89$ GHz. The known existence of modes at THz frequencies for ILs as well as for AN is confirmed by comparison of the 'infinite-frequency' permittivity determined from data at $\nu \leq 89$ GHz and the square of the refractive index measured at the sodium-D line, n_{589}^2, which may be assumed as the high-frequency limit of $\varepsilon'(\nu)$: generally, $n_{589}^2 < \varepsilon_\infty$ is found for [emim][EtSO$_4$] + AN mixtures (compare Table 5.7 and Appendix A.1), indicating that modes are present at THz frequencies.

Low-frequency mode. No literature data for the viscosities of mixtures other than [bmim]-[BF$_4$], [bmim][PF$_6$] and [emim][EtSO$_4$] + AN appear to be available. However, for the latter mixtures the increase of τ_1 at $x_{\text{IL}} \gtrsim 0.3$ correlates with the increase in viscosity (obtained by interpolation of data given in refs. 187 & 194 and Appendix A.1) of these mixtures, as indicated by the linearity of a plot of the corresponding molecular relaxation time τ_1' (via Eq. 1.80) against viscosity (Figure 5.13). As has been found previously for various neat ILs and for IL + DCM mixtures,[36–38,48] quantitative analysis of τ_1' on the basis of Eq. 1.77 yields unreasonably small values of $V_{\text{eff},1} = 0.93$ Å^3 for [bmim][BF$_4$], $V_{\text{eff},1} = 1.1$ Å^3 for [bmim][PF$_6$] and $V_{\text{eff},1} = 0.93$ Å^3 for [emim][EtSO$_4$]. However, a recent combined dielectric and optical Kerr-effect study showed anisotropic reorientation of the cations through large-angle jumps,[39] which suggests that the SED theory is not applicable to the relaxation phenomena of ILs. Nevertheless, the variation of τ_1 at $x_{\text{IL}} \gtrsim 0.3$ (Figure 5.7c) together with the asymmetry of the $V^{\text{E}} = f(x_{\text{IL}})$ curves (Figure 5.4) show that the present ILs keep their molten-salt character up to high dilutions with AN. In essence, co-solvent additions 'lubricate' the IL dynamics, which become more homogenous, consistent with the smooth increase of the CC broadness parameter, α, at $x_{\text{IL}} \gtrsim 0.3$ (Figure 5.7d). Moreover, α depends on the cations in the order [emim]$^+$ < [bmim]$^+$ < [hmim]$^+$ (Table 5.6). As discussed elsewhere,[37] the molecular interpretation of α values is not straightforward, but their increase with increasing x_{IL} probably reflects a more heterogenous molecular-level environment, being experienced by the cations. A combined DR and OHD-RIKE study,[39] and various other techniques,[139,199,200] have confirmed the presence of locally-heterogenous environments in ILs, originally suggested by computer simulations.[135,136,179,201–203] Among the latter, the picture presented by Canongia Lopes et al.[135] is particularly informative. Their view suggests that in ILs, polar domains form a 3D network, which is permeated by increasingly larger non-polar domains (ultimately producing microphase segregation) with increasing cation side-chain length.

Figure 5.8 compares (for [bmim][BF$_4$] + AN mixtures) the observed amplitude of relaxation process 1, S_1, with $S_{\text{[bmim]}^+}$, the amplitude estimated from Eq. 1.68 assuming that all the cations present contribute to the spectrum and that their effective dipole moment, $\mu_{\text{eff},+}$, has

5.2. DIELECTRIC PROPERTIES

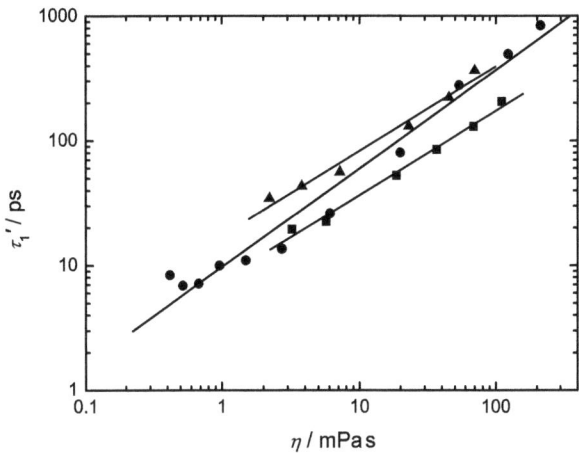

Figure 5.13: Microscopic relaxation time of process 1, τ_1', of binary IL + AN mixtures at 25 °C plotted as function of solution viscosity for the ILs: ■, [bmim][BF$_4$]; ●, [bmim][PF$_6$]; ▲, [emim][EtSO$_4$]. The lines are linear fits to the experimental data shown ($x_{IL} \gtrsim 0.3$).

the same value as in the neat IL (for [bmim][Cl] + AN mixtures, the value for $\mu_{\text{eff},+}$ obtained from [bmim][BF$_4$] was used in the calculations). Note, that the variation of S_1 with respect to concentration is similar for all IL + AN mixtures studied, except for [emim][EtSO$_4$] + AN, which will be discussed below. The experimental values of S_1 clearly cannot be explained by assuming only the presence of the dipolar [bmim]$^+$ ions. Nor would ion pairs (see below) be expected to contribute significantly, especially at low x_{IL} where $S_1 \gg S_{[\text{bmim}]^+}$ (Figure 5.8). As discussed above in terms of solvation numbers, the contribution of bulk-like AN to the high-frequency process is less than expected and it vanishes with rising x_{IL}. As suggested by the relatively modest solvating properties of AN, interactions of AN with the cations would be expected to be relatively weak. This means that 'irrotational bonding', a 'freezing out' (immobilization on the time scale of DR spectroscopy) of AN molecules by the IL, would be unlikely in the present mixtures. Strong interactions of [bmim]$^+$ with AN can also be excluded on the basis of UV absorption spectra.[204] Nevertheless, recent MD simulations have shown that AN interacts to some extent with both the charged and the non-polar domains present in ILs.[32,33,135] It is plausible to assume that the dynamics of solvating AN molecules are slowed

compared to 'bulk'-like AN and thus contribute to the low-frequency process.
Provided that the contribution of bulk AN to S_2 is dominant at low x_{IL} but negligible at high x_{IL}, the amplitude of process 1 can be reduced by the contribution of these 'slow' AN molecules to yield:

$$S_{1,\text{red}} = \begin{cases} S_1 - (S_{\text{AN}} - S_2) & \text{for } x_{\text{IL}} \lesssim 0.3, \\ S_1 - S_{\text{AN}} & \text{for } x_{\text{IL}} \gtrsim 0.3 \end{cases} \quad (5.9)$$

Figure 5.8 shows the resulting amplitude, which only contains the contributions of cations and ion pairs. As expected, it coincides with the cation amplitude, $S_{[\text{bmim}]^+}$, at high concentrations ($x_{\text{IL}} \gtrsim 0.3$). The higher amplitudes at $x_{\text{IL}} \lesssim 0.2$ reflect the presence of IPs.

Figure 5.9a shows for [emim][EtSO$_4$] + AN mixtures the observed amplitude of relaxation process 1, S_1. Compared to other IL + AN mixtures (see above), the variation of S_1 with IL content is rather different: it increases smoothly and does not show a maximum. Moreover, at low x_{IL}, the values of S_1 are well below the corresponding ones in IL + AN mixtures discussed above. This indicates, that a contribution due to 'slow' AN molecules can be excluded for process 1 in [emim][EtSO$_4$] + AN mixtures. Note, that due to orientational correlations among the cations (see below), a comparison of S_1 with the 'expected' $S_{[\text{emim}]^+}$ is not advisable.

Anion process. Figure 5.9b compares for [emim][EtSO$_4$] + AN mixtures the observed amplitude of the anion process, S_{anion}, with $S_{\text{EtSO}_4^-}$, again calculated from Eq. 1.68 assuming that all anions present contribute to the spectrum and that their effective dipole moments have the same values as in the neat IL. Note, that for neat [emim][EtSO$_4$], dipole correlations among the anions are pronounced, yielding $g_- = 0.024$ at 25 °C.[37] For [emim][EtSO$_4$] + DCM mixtures it was found that g_- varies with IL content,[37] but the comparison in Figure 5.9b is made assuming $g_-(x_{\text{IL}}) = 0.024$.

The variation of S_{anion} with IL content is similar to the results obtained for [emim][EtSO$_4$] + DCM mixtures. This indicates, that as for [emim][EtSO$_4$] + DCM mixtures orientational correlations among the anions may be present, particularly at high x_{IL}, where S_{anion} ultimately reaches the value of $\sim$2.2 for the neat IL. However, the absolute values of S_{anion} are much higher for [emim][EtSO$_4$] + AN mixtures, suggesting the presence of 'slow' AN relaxation in the anion process. Calculation of the corresponding reduced anion amplitude, $S_{\text{anion,red}}$, following Eq. 5.9, yields the values shown in Figure 5.9b. These results again suggest the existence of 'slow' AN molecules, but the high scatter of the data does not permit an unambiguous identification of that species nor the calculation of g_-.

Additional information comes from the relaxation times of the anion process, τ_{anion}. For low x_{IL}, the values for τ_{anion} increase with increasing IL content, but they are well below the correspond-

5.2. DIELECTRIC PROPERTIES

ing [emim][EtSO$_4$] + DCM results. This is in contrast to the high similarity of the viscosities for neat AN and DCM (η(DCM, 25 °C) = 0.423 mPa s and η(AN, 25 °C) = 0.341 mPa s),[120] and for [emim][EtSO$_4$] + AN and + DCM mixtures (compare values given in Appendix A.1 and ref. 187). Quantitative agreement would not be expected, as the SED theory was shown to be inapplicable to ILs,[39] but a scaling of the dielectric properties to the mixture viscosity has been shown in the present work. Hence, the strong decrease of τ_{anion} with decreasing IL content is most probably the result of the superposition of 'slow' AN and anion relaxations.

To summarize, the formal description of the DR spectra obtained for IL + AN mixtures using the CC + D model can be interpreted as follows: for the low-frequency process, cation jump reorientation contributes over the whole composition range, superimposed by IP contributions at $x_{\text{IL}} \lesssim 0.3$. The high-frequency process is dominated by contributions from fast IL modes (librations and intermolecular vibrations) at $x_{\text{IL}} \gtrsim 0.8$, whereas the major part of that process is due to reorientation of bulk-like AN molecules at lower IL content. Note, that for the present IL + AN mixtures, a separate anion process is not observable in the spectra at $\nu \leq 89$ GHz for anions having a zero or low dipole moment. Recent DR studies of neat ILs containing the anions BF$_4^-$, PF$_6^-$, DCA$^-$ and NTf$_2^-$ have indeed shown that a process due to anion reorientation is not present (or below detection limit) in the MW spectra.[38,39] However, for the highly dipolar anion in [emim][EtSO$_4$] + AN mixtures, a separate anion process is present, in concordance to the neat IL and to [emim][EtSO$_4$] + DCM mixtures.[37] A contribution of 'slow' AN is suggested by the analysis of the present spectra. Depending on the species present, this relaxation is superimposed by cation or anion modes.

Ion pairing. The strong decrease of both τ_1 and α at low x_{IL} in IL + AN mixtures (Figures 5.7c and 5.7d) is characteristic for the coexistence of ion pairs and cations. Similar curves have been observed in IL + DCM mixtures and have been assigned to contact ion pair (CIP) formation.[37,48] Dielectric spectroscopy is particularly sensitive to the presence of ion pairs in solution.[34] Relaxation modes for IPs are expected in the high MHz- to low GHz-regions, corresponding to relaxation times in the vicinity of 1 ns. Among the possible species, contact ion pairs (CIPs), solvent-shared ion pairs (SIPs) or double solvent separated ion pairs (2SIPs), can contribute to dielectric spectra.[43]

It is possible to use Eq. 1.68 with the reduced relaxation amplitude, $S_{1,\text{red}}$, to calculate the effective molecular dipole moment, $\mu_{\text{eff},1,\text{red}}$, of the species contributing to that amplitude. For all IL + AN mixtures, the values obtained (with data for [bmim][BF$_4$] + AN mixtures being presented in Figure 5.14) show a pronounced decrease with increasing x_{IL}. For ILs having anions with zero or low dipole moment, $\mu_{\text{eff},1,\text{red}}$ ultimately reaches the value of $\sim$4.4 D at $x_{\text{IL}} \gtrsim 0.2$, which is practically the value found for neat [bmim][BF$_4$], $\mu_{\text{eff},+} = (4.4 \pm 0.9)\,\text{D}$.[38] This is in concordance with the center of mass dipole moment, $\mu_{\text{calc},+} = (4.9\pm0.7)\,\text{D}$,[205] obtained via DFT calculations on the B3LYP/6-31G(d,p) level. Similar curves to that shown in Figure 5.14 were obtained for [bmim][BF$_4$] + DCM mixtures and were attributed to the formation of CIPs.[48] Various experimental techniques have been used to prove the existence of ion pairs in IL + solvent mixtures.[25,27–29,35,56,206,207] Typically, the degree of association is most pronounced at low IL contents ($x_{\text{IL}} \lesssim 0.1$). Note that at *very* low x_{IL} Le Chatelier's principle demands that the IPs become fully dissociated, as in any electrolyte solution.

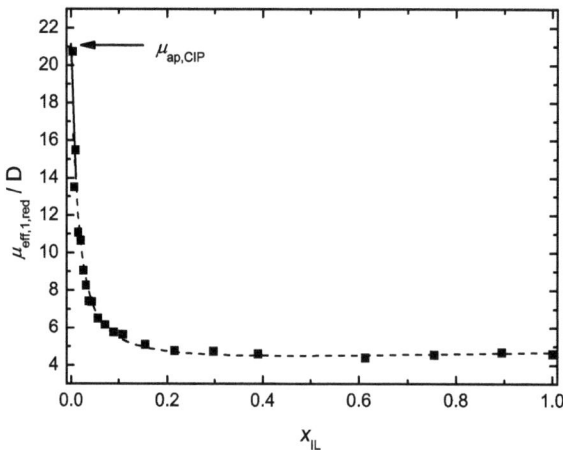

Figure 5.14: Effective dipole moments, $\mu_{\text{eff,1,red}}$, including contributions from ion pairs and cations, for [bmim][BF$_4$] + AN mixtures at 25 °C. The value of the apparent dipole moment of the ion pair, $\mu_{\text{ap,CIP}}$, is indicated.

Since in the neat IL, $\mu_{\text{eff},+} \approx \mu_{\text{calc},+}$, it follows that neither anions (as would be expected from their zero or small dipole moments) nor IPs (for which $\mu \neq 0$) make a significant contribution. Furthermore, as $\mu_{\text{eff,1,red}} \approx \mu_{\text{eff},+}$ at $x_{\text{IL}} \gtrsim 0.2$, the mixtures in this composition range must be essentially 'IL-like' and orientational correlations among the ions must be absent. At $x_{\text{IL}} \lesssim 0.2$, $\mu_{\text{eff,1,red}}$ increases dramatically because of the formation of IPs and thus the dipole moment of the ion pairs, $\mu_{\text{ap,CIP}}$, can be approximated as $\mu_{\text{eff,1,red}}(x_{\text{IL}} \to 0)$, as orientational correlations between ion pairs can be neglected ($g_{CIP} = 1$ in Eq. 1.71). Note, that for [emim][EtSO$_4$] + AN mixtures, the smooth increase of $\mu_{\text{eff},1}$ (which corresponds to $\mu_{\text{eff},1,\text{red}}$ in IL + AN mixtures discussed above) at $x_{IL} \gtrsim 0.1$ (Figure 5.15) indicates dipole-dipole correlations among the cations. This is in concordance with a previous investigation of [emim][EtSO$_4$] + DCM mixtures.[37] In addition to $\mu_{\text{ap,CIP}}$, the value of the apparent dipole moment of the cation, $\mu_{\text{ap},+} = 6.53\,\text{D}$, can be estimated by linear extrapolation of $\mu_{\text{eff},1}$ to infinite dilution of the IL in AN. The value is somewhat higher compared to the result from [emim][EtSO$_4$] + DCM mixtures ($\mu_{\text{ap},+} = 4.64\,\text{D}$),[37] although the reasons for the difference are unknown at present. Possibly, the variation of $\mu_{\text{eff},+}$ vs. x_{IL} is not linear at very low x_{IL} and consequently, the extrapolation performed herein would not be justified. However, using Eq. 1.71, the correlation factor g_+ of the cation can be determined. The values obtained show a smooth increase from $g_+ = 1$ at infinite dilution to $g_+ = 1.3$ for neat [emim][EtSO$_4$], assuming that a linear extrapolation to $x_{IL} = 0$ is possible. Whereas the increase of g_+ is consistent with the results from [emim][EtSO$_4$] + DCM mixtures, the absolute values cannot be compared, as they strongly depend on the value of $\mu_{\text{ap},+}$. In principle, this procedure could be applied to the anion process in [emim][EtSO$_4$] + AN mixtures, but due to the likely contribution of 'slow' AN molecules, the analysis in terms of the corresponding $\mu_{\text{ap},-}$ and g_- is not advisable.

The estimate of $\mu_{\text{ap,CIP}} = 21.1\,\text{D}$ for a [bmim][BF$_4$] contact ion pair (by linear extrapolation of low-concentration values of $\mu_{\text{eff},1,\text{red}}$ as function of x_{IL}) is broadly consistent with semi-empirical MOPAC[208] calculations of $\mu_{\text{ap,CIP}} \approx (18.6 - 23.1)\,\text{D}$ for various conformers (see Table 5.9 for all IL + AN mixtures studied), and with the 'experimental' values for CIPs of [bmim][BF$_4$] in DCM ($\mu_{\text{ap,CIP}} = 22\,\text{D}$).[48] The same is found for [emim][EtSO$_4$] in AN, where the value of $\mu_{\text{ap,CIP}} = 17.4\,\text{D}$ compares quite well with the results from [emim][EtSO$_4$] + DCM mixtures ($\mu_{\text{ap,CIP}} = 22.9\,\text{D}$).[37] MOPAC[208] calculations of SIP and 2SIP formed by [bmim][BF$_4$] and AN yielded much higher values for the apparent dipole moments ($\mu_{\text{ap,SIP}} = 36.5\,\text{D}$ and $\mu_{\text{ap,2SIP}} = 40.7\,\text{D}$). Thus, together with the estimates of the volumes of rotation (see below), the species present are undoubtedly contact ion pairs.

If only CIPs and free ions are present at low IL concentrations, Eq. 1.84 (where $g_j = 1$ and thus $\mu_{\text{eff},j} = \mu_{\text{ap},j}$ is assumed, except for [emim][EtSO$_4$] + AN mixtures, where g_+ was explicitly calculated) can be used to determine the molar concentrations of the contributing species. The results so obtained for c_{CIP} are plotted as relative concentrations, c_{CIP}/c_{IL}, in Figure 5.16, together with the association constants, K_A (Eq. 1.86, with $c = c_{IL}$) for mixtures of [emim][BF$_4$], [bmim][BF$_4$] and [hmim][BF$_4$] with AN, as well as mixtures of [emim][EtSO$_4$] with AN or DCM. The values of K_A can be used to estimate the standard (infinite dilution) association constant for the formation of the CIPs, K_A°, by extrapolation with Eq. 1.87. For AN at 25 °C, the Debye-

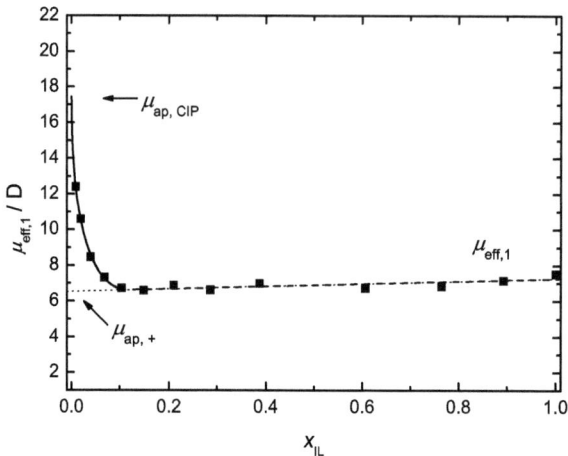

Figure 5.15: Effective dipole moments, $\mu_{\text{eff},1}$, including contributions from ion pairs and cations, for [emim][EtSO$_4$] + AN mixtures at 25 °C. The values of the apparent dipole moment of the ion pair, $\mu_{\text{ap,CIP}}$, and the cation, $\mu_{\text{ap},+}$, are indicated.

Hückel coefficients are $A_{\text{DH}} = 1.643\,\text{L}^{1/2}\,\text{mol}^{-1/2}$ and $B_{\text{DH}} = 4.857 \cdot 10^9\,\text{L}^{1/2}\,\text{mol}^{-1/2}\,\text{m}^{-1}$. For variation of the cations in ILs the extrapolated values of log K_A° follow the order [emim][BF$_4$] > [bmim][BF$_4$] > [hmim][BF$_4$] (Table 5.10). Regarding combinations of the [bmim]$^+$ cation with various anions, one finds for log K_A°: [bmim][Cl] > [bmim][DCA] > [bmim][BF$_4$] > [bmim][PF$_6$]. This is consistent with the decreasing charge density of the anions, resulting in weakened interactions with increasing size of the ions. The same cation dependence was found for these ILs dissolved in DCM[209] and similar values have been determined for CIPs of 1,1-electrolytes in AN solutions.[198,210] Consistent with the bulk dielectric permittivities of AN and DCM, association of the IL is much higher in the latter.

Attempts to separate the ion pair amplitude from process 1 by fitting an additional Debye equation at low frequency were successful at low IL concentrations for some of the mixtures, but in most cases this required fixing of parameters. Table 5.11 lists the parameters so obtained for the corresponding D + CC + D model. Due to strong mode overlap, the splitting of the CIP mode is only possible at *very* low concentrations and the parameters show considerable

5.2. DIELECTRIC PROPERTIES

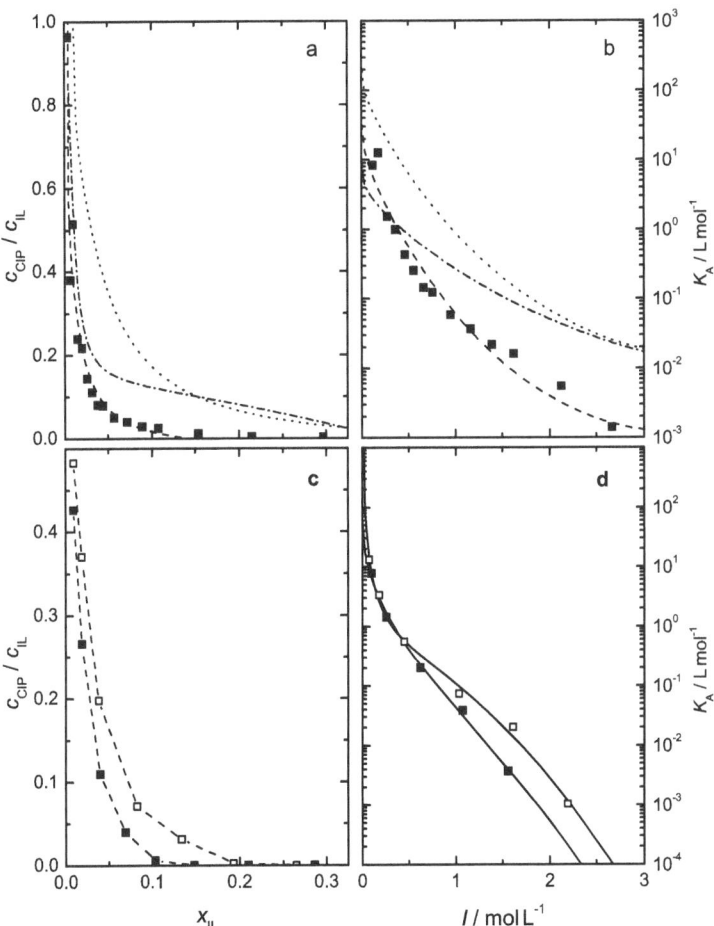

Figure 5.16: (a) Relative contact ion pair concentrations, $c_{\text{CIP}}/c_{\text{IL}}$, together with (b) association constants, K_A, of [emim][BF$_4$] (dotted line), [bmim][BF$_4$] (dashed line) and [hmim][BF$_4$] (dash-dotted line) in IL + AN mixtures at 25 °C. (c), (d) Corresponding data for [emim][EtSO$_4$] + AN (full symbols) and [emim][EtSO$_4$] + DCM (open symbols).[37] For visual clarity, some of the experimental data are omitted. Lines in (b) and (d) represent fits according to Eq. 1.87. Note that $c_{\text{CIP}}/c_{\text{IL}}$ must $\to 0$ as $x_{\text{IL}} \to 0$.

Table 5.9: Values of the Apparent Dipole Moment, $\mu_{\text{ap,CIP}}$, Obtained by Extrapolation of Experimental Data and by MOPAC[208] Calculations for Contact Ion Pairs (CIPs) in IL + AN mixtures at 25 °C.

CIP	$\mu_{\text{ap,CIP}}$/ D experiment	$\mu_{\text{ap,CIP}}$/ D MOPAC calculation[a]
[emim][BF$_4$]	11.1	17.7 − 19.8
[bmim][BF$_4$]	21.1	18.6 − 23.1
[hmim][BF$_4$]	15.9	18.4 − 26.3
[bmim][PF$_6$]	15.1	19.0 − 22.6
[bmim][Cl]	15.6	14.6 − 18.6
[bmim][DCA]	14.7	20.0 − 22.3
[hmim][NTf$_2$]	15.5	19.9 − 26.7
[emim][EtSO$_4$]	17.4	16.6 − 19.7

[a] Range given refers to various conformers.

scatter. Generally, the splitting was not possible at $x_{\text{IL}} \gtrsim 0.15$. However, Eqs. 1.86 & 1.87 applied to c_{CIP} directly calculated from S_{CIP} of the D + CC + D fit yielded the parameters listed in Table 5.10. Comparison of the values of $\log K_{\text{A}}^{\circ}$ of the CC + D and the D + CC + D models show in general good agreement. Moreover, $S_{1,\text{red}}$ derived from the D + CC + D model via Eq. 5.9 coincides with $S_{\text{[bmim]}^+}$ at low x_{IL} (Figure 5.8). Thus, the applied splitting of the CC mode is self-consistent and the parameter fixing seems to be justified.

Clearly, the present values of $\log K_{\text{A}}^{\circ}$ (Table 5.10) for ILs in AN are only approximate. Precise measurements of the molar conductivity of dilute solutions are more suitable to determine association constants of 'symmetric electrolyte' ILs in AN.[§] The values so obtained (Table 5.10) agree quite well with the results of the DR studies. Most importantly, general trends in $\log K_{\text{A}}^{\circ}$, for example high and low degrees of association for [bmim][Cl] and [hmim][NTf$_2$], respectively, are quite well reflected by the conductivity measurements. Quantitative agreement with DR studies cannot be expected due to the uncertainties in the determination of the apparent dipole moment. This can be shown by the following example: increasing the value of $\mu_{\text{ap,CIP}}$ given in Table 5.9 for [bmim][BF$_4$] by a factor of 1.25, gives $\log K_{\text{A}}^{\circ} = 1.18$, which is in perfect agreement to the result of the conductivity measurement. On the other hand, the values of

[§]We thank Prof. Dr. Marija Bešter Rogač, who performed the measurements and the data analysis applying Barthel's low-concentration chemical model (lcCM).[185]

5.2. DIELECTRIC PROPERTIES

Table 5.10: Standard Association Constants, $\log K_A^\circ$, and Parameters A_K and B_K (Eq. 1.87) of IL + AN Mixtures at 25 °C Derived from DR Studies (CC + D and D + CC + D Models) and High Precision Conductivity Measurements.[a]

IL	CC + D			D + CC + D			Cond.[b]
	$\log K_A^\circ$	A_K $\overline{\text{L mol}^{-1}}$	B_K $\overline{\text{L}^{3/2}\text{mol}^{-3/2}}$	$\log K_A^\circ$	A_K $\overline{\text{L mol}^{-1}}$	B_K $\overline{\text{L}^{3/2}\text{mol}^{-3/2}}$	$\log K_A^\circ$
[emim][BF$_4$]	2.39	-2.96	1.02	2.30	-3.59	1.32	1.201
[bmim][BF$_4$]	1.53	-3.53	1.29	1.71	-6.04	3.25	1.193
[hmim][BF$_4$]	0.99	-1.50	0.436	1.30	-2.61	1.02	1.217
[bmim][PF$_6$]	1.31	-2.91	1.16	–	–	–	1.193
[bmim][Cl]	2.85	-6.21	3.60	2.27	-4.56	2.50	1.804
[bmim][DCA]	2.34	-4.06	1.61	1.33	-2.75	1.07	–[c]
[hmim][NTf$_2$]	0.57	-1.41	0.407	–	–	–	1.103
[emim][EtSO$_4$]	1.78	-0.309	1.60	–	–	–	1.615

[a] Analysis performed for mixtures where sufficient data available.
[b] Provided by M. Bešter Rogač.§
[c] Not available.

$\log K_A^\circ = 2.85$ for [bmim][BF$_4$] and 2.83 for [bmim][PF$_6$] in AN, determined by Wang et al.[211] using low precision conductivity data, are almost certainly too high. No further literature data appear to be available for the association of the present ILs in AN.

The ILs are only moderately associated at high dilutions in AN (Table 5.10). Even so, the relative concentrations of IPs in these solutions become quite high at low x_{IL} (Figure 5.16a). As already stated, the higher degree of association of [emim][BF$_4$] compared with [bmim][BF$_4$] (Figure 5.16) is in accord with the decreasing charge density of the cations. The values of K_A for [hmim][BF$_4$] in AN show a rather different dependence on I compared to [emim][BF$_4$] and [bmim][BF$_4$] (Figure 5.16b). One could imagine the formation of larger, polar aggregates, that form at higher IL concentrations and contribute to process 1, as long as they rotate on the experimentally accessible timescale ($\tau \lesssim 1$ ns). This picture is consistent with MD simulations,[135] which predict enhanced domain growth with increasing length of the cation side chain. For [bmim][BF$_4$] + AN mixtures, Brillouin light scattering excluded any structural coordination on the timescale $\lesssim 1$ ns.[30] However, the influence of the alkyl chain is still unclear, as the results of the conductivity measurements show only little dependence of $\log K_A^\circ$ for a given anion and varying cation. The balance of all ion/ion, ion/solvent and solvent/solvent interactions determines the association behavior of the IL in solution.

Table 5.11: Fit Parameters of Eq. 1.61 for the Observed Dielectric Spectra of Dilute Mixtures of ILs with AN at 25 °C Assuming the D + CC + D Model: Static Permittivities, ε; Relaxation Times, τ_j, and Amplitudes, S_j, of Process j and of the Contact Ion Pair Process (CIP); Cole-Cole Shape Parameter, α, of the Cation Process; Infinite Frequency Permittivity, ε_∞, and Reduced Error Function of the Overall Fit, χ_r^2.[a]

x_{IL}	ε	S_{CIP}	τ_{CIP}/ps	S_1	τ_1/ps	α	S_2	τ_2/ps	ε_∞	$\chi_r^2/10^{-4}$
0[a]	35.84						32.5	3.32	3.33	
				[emim][BF$_4$]						
0.01082	36.72	1.99	90.0	2.27	9.23	0.05[b]	28.8	3.36	3.65	116
0.01646	36.75	2.26	78.4	3.46	9.84	0.05[b]	27.3	3.35	3.75	172
0.02252	36.67	1.96	85.7	4.10	14.8	0.05[b]	27.4	3.30[b]	3.17	366
0.04929	35.90	1.89	82.6	7.86	13.8	0.05[b]	22.3	3.48	3.83	213
0.08059	34.71	1.41	121	9.24	16.3	0.05[b]	18.9	4.22	5.11	199
0.1218	32.63	1.33	87.9	12.0	15.0	0.05[b]	14.7	4.08	4.62	142
0.1736	30.40	1.05	98.0	14.4	15.4	0.06[b]	10.4[b]	3.94	4.58	131
				[bmim][BF$_4$]						
0.003131	36.49	1.03	206	2.47	7.76	0.05[b]	30.2	3.11	2.81	108
0.006352	36.65	1.32	161	1.41	18.2	0.05[b]	30.1	3.38	3.84	157
0.009410	36.43	1.73	96.3	2.57	9.93	0.042	29.1	3.22	3.04	135
0.01472	36.57	1.56	114	3.82	14.8	0.05[b]	28.1	3.20	3.13	329
0.01981	36.02	2.27	65.3	4.76	9.69	0.05[b]	25.7	3.24	3.25	205
0.02574	36.21	1.76	90.4	6.56	12.5	0.05[b]	24.5	3.17	3.44	447
0.03142	35.52	2.60	55.9	6.14	10.5	0.05[b]	23.5	3.27	3.32	219
0.03774	35.69	1.03	128	6.03	20.4	0.05[b]	24.5	3.69	4.10	612
0.04389	35.26	1.60	94.7	6.54	17.6	0.035	23.0	3.68	4.09	284
0.05668	34.78	1.13	117	8.20	17.6	0.05[b]	20.8	3.88	4.62	586
0.07179	33.54	1.28	70[b]	10.4	15.1	0.10	17.7	3.87	4.12	178
0.08907	33.11	1.00	119	12.0	15.8	0.07[b]	15.6	3.85	4.59	554
0.1077	31.89	1.08	88.9	12.2	16.8	0.087	14.1	4.16	4.55	226

[a] Parameters from Section 3.2.
[b] Parameter fixed during fitting procedure.

5.2. DIELECTRIC PROPERTIES

Table 5.11 Continued.

x_{IL}	ε	S_{CIP}	τ_{CIP}/ps	S_1	τ_1/ps	α	S_2	τ_2/ps	ε_∞	$\chi_r^2/10^{-4}$
				[hmim][BF$_4$]						
0.004117	36.42	1.36	126	1.40	7.83	0.05^b	30.1	3.30^b	3.6	165
0.008445	36.49	1.94	97.0	4.55	7.08	0.05^b	26.8^b	3.13	3.2	125
0.01744	36.23	2.28	79.0	3.67	13.4	0.05^b	27.1	3.30	3.1	213
0.03816	35.24	2.35	77.8	5.20	18.7	0.05^b	23.3	3.90	4.4	270
0.06486	33.87	1.92	104	8.85	19.3	0.07^b	19.2	3.86	3.9	191
0.09763	31.66	3.49	65.5	11.2	13.4	0.1^b	12.8	3.99	4.2	99.0
0.1399	29.85	2.63	109	13.1	17.8	0.1^b	9.77	4.23	4.4	127
				[bmim][PF$_6$]						
0.007570	36.22	0.983	150^b	1.22	50^b	0.05^b	30.8	3.35	3.21	270
0.01576	36.07	0.874	400	2.60	42.5	0.05^b	28.8	3.60	3.78	448
				[bmim][Cl]						
0.005508	36.99	0.168	141	1.71	62.3	0.05^b	31.2	3.5^b	3.90	436
0.01261	37.97	2.87	71.5	2.77	11.4	0.05^b	28.7	3.39	3.58	243
0.02572	39.12	4.19	77.1	4.43	15.7	0.05^b	26.6	3.51	3.89	220
0.03837	39.44	5.30	73.1	5.27	15.4	0.05^b	24.6	3.76	4.31	380
0.05100	39.45	6.01	76.0	7.60	13.6	0.05^b	21.9	3.59	3.98	294
0.07585	38.81	6.75	81.2	9.16	14.1	0.05^b	18.3	3.96	4.65	412
				[bmim][DCA]						
0.01992	36.31	1.96	88.9	2.56	19.7	0.05^b	27.3	3.74	4.46	277
0.04752	34.73	1.85	83.0	6.62	15.8	0.05^b	22.4	3.70	3.89	293
0.07852	32.80	2.22	76.4	10.7	12.3	0.05^b	15.6	3.70	4.20	157
0.1175	30.59	1.90	110	10.1	16.4	0.05^b	14.0	4.31	4.55	135
				[hmim][NTf$_2$]						
0.002346	36.47	0.883	389	0.666	50^b	0.1^b	31.3	3.36	3.59	188

^{a} Parameters from Section 3.2.
^{b} Parameter fixed during fitting procedure.

As the present ILs seem to behave as more-or-less as conventional electrolytes in dilute solution in AN it is useful to analyse their behavior in this region in more detail. From all the mixtures studied, the following analysis could not be performed for [bmim][PF$_6$], [hmim][NTf$_2$] and [emim][EtSO$_4$] + AN mixtures, as the fitting of the additional low-frequency process was not possible. On the basis of the equilibrium of cations, anions and ion pairs depicted in Eq. 1.82, the decrease of τ_{CIP} from the D + CC + D model as a function of (electrolyte) concentration (Eq. 1.95) is related to two processes: the reorientation of dipolar ion pairs and the chemical equilibrium of the species present. When K_A° and c_{CIP} are known (see above), a linear fit of τ_{CIP}^{-1} vs. $(c_{\text{IL}} - c_{\text{CIP}})$ yields quantitative information on the ion pair formation and decay kinetics. Generally, the plots were found to be linearly increasing only at $x_{\text{IL}} < 0.1$ with a following constant part (within the scatter of the data) at higher concentrations. Figure 5.17 shows typical results for [bmim][BF$_4$] + AN mixtures. This can be interpreted in the following way: at very low concentrations of IL in AN, the chemical equilibrium (Eq. 1.82) has a major impact on the relaxation rate of the ion pairs, whereas with increasing c_{IL} only the reorientation of the ion pair and thus τ'_{CIP} affect the overall relaxation rate.

The rate constants for CIP formation and decay given in Table 5.12 were determined from the linear parts of the plots and have to be regarded as estimates. Nevertheless, for [bmim][BF$_4$] in AN, the rate constant for ion pair formation agrees well with the value calculated via Eigen's theory (Eqs. 1.96-1.98, λ_j^∞ provided by M. Bešter Rogač[§]) yielding $k_1^{\text{D}} = 2.0 \cdot 10^{10}\,\text{L}\,\text{mol}^{-1}\,\text{s}^{-1}$, whereas our value for the rate constant for ion pair decay is smaller than the predicted one, $k_{-1}^{\text{D}} = 2.2 \cdot 10^{10}\,\text{s}^{-1}$. Quantitative agreement cannot be expected for various reasons: non-Coulombic interactions surely play a significant role in the ion pair formation process, the Eigen theory[91] is only valid for infinite dilution of the ions and the strong mode overlap leads to high uncertainties in the present analysis. Nevertheless, the comparison suggests diffusion controlled ion pair formation, but not for the decay process. The values obtained for the ratios $k_1/k_{-1} \approx 10^2\,\text{L}\,\text{mol}^{-1}$ can be rationalized by the values obtained for $\log K_A^\circ \approx 2$ (Table 5.10). Use of Eqs. 1.77 & 1.79 to estimate the values of $V_{\text{eff,CIP}}$ yield small values compared to those derived from MOPAC[208] calculations (Table 5.12), but they are still higher than the van der Waals volumes obtained from semiempirical calculations ($\sim$200 Å^3 for the present ILs).[208] For [bmim][BF$_4$] + AN the determination of $V_{\text{eff,CIP}}$ should be most reliable due to the large number of mixtures studied at low x_{IL}. Interestingly, $V_{\text{eff,CIP}}$ compares quite well with the corresponding MOPAC estimate (Table 5.12). Taking into account the uncertainties of the present analysis, the results clearly indicate the formation of contact ion pairs at low IL concentrations.

5.2.2 IL + methanol mixtures

Fit model

The dielectric spectra for [bmim][BF$_4$] + MeOH mixtures showed, similarly to IL + AN mixtures, a smooth transition from neat MeOH to the neat IL. The retrieval of the 'true' relaxation mechanism in these mixtures is, however, even more problematic. With the knowledge gained

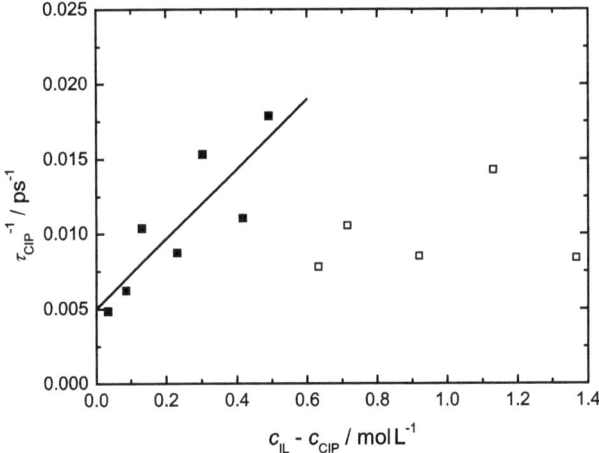

Figure 5.17: Relaxation rate of the contact ion pairs, τ_{CIP}^{-1}, plotted as function of ionic strength, $c_{\text{IL}} - c_{\text{CIP}}$, for [bmim][BF$_4$] + AN mixtures at 25 °C. The line is a linear fit, according to Eq. 1.95, to the data presented as full symbols.

Table 5.12: Orientational Relaxation Rates, τ'_{CIP}, and Rate Constants of Contact Ion Pair Formation, k_1, and Decay, k_{-1}, Together with Volumes of Rotation Derived from Experimental Data, $V_{\text{eff,CIP}}$, and MOPAC[208] Calculations, $V_{\text{rot,CIP}}$, for ILs in AN at 25 °C.[a]

IL	k_1	k_{-1}	τ'_{CIP}	$V_{\text{eff,CIP}}$	$V_{\text{rot,CIP}}$
	$10^{10}\,\text{L}\,\text{mol}^{-1}\,\text{s}^{-1}$	$10^8\,\text{s}^{-1}$	ps	Å^3	Å^3
[emim][BF$_4$]	0.0028	0.0014	85.0	342	570
[bmim][BF$_4$]	1.2	2.3	209	841	964
[hmim][BF$_4$]	0.39	1.9	110	442	1348
[bmim][PF$_6$]	–	–	–	–	–
[bmim][Cl]	0.76	0.41	112	452	843
[bmim][DCA]	0.11	0.49	94.3	379	1089
[hmim][NTf$_2$]	–	–	–	–	–
[emim][EtSO$_4$]	–	–	–	–	–

[a] Analysis performed for mixtures where sufficient data available.

from the dielectric relaxation of the neat components (Section 3), various modes are expected in the frequency range under investigation ($0.2 \leq \nu/\text{GHz} \leq 89$). An extensive analysis using Eq. 1.61 indicated that, depending on the composition of the mixture, up to four processes are present (Table 5.13): at $x_{\text{IL}} \leq 0.1242$, a combination of four Debye equations, the D + D + D + D model, fit the spectra best. At higher x_{IL}, the first process is broadened, leading to the CC + D + D + D model. For the highest IL content studied ($x_{\text{IL}} = 0.5583$), the CC + D + D model was sufficient to fit the spectrum. Note, that due to strong mode overlap and the complexity of the DR spectra for these mixtures, the scatter of the relaxation parameters with respect to composition is high (see below). Thus, mixtures were only studied for $x_{\text{IL}} \leq 0.5583$ and the following discussion will have more qualitative rather than quantitative character.

Composition dependence of relaxation parameters

In the present [bmim][BF$_4$] + MeOH mixtures at low x_{IL}, the relaxation times of the two highest-frequency Debye modes, here labelled processes 3 and 4, are close to the values found in neat MeOH ($\tau_3 = 8.23\,\text{ps}$ and $\tau_4 = 1.05\,\text{ps}$) and practically do not depend on composition. In contrast with process 3, the highest-frequency process is also observable in the neat IL. This indicates, as for IL + AN mixtures, that process 4 is a superposition of the relaxations present

5.2. DIELECTRIC PROPERTIES

Table 5.13: Fit Parameters of Eq. 1.61 for the Observed Dielectric Spectra of Mixtures of [bmim][BF$_4$] with MeOH at 25 °C Assuming the CC + D + D + D Model: Static Permittivities, ε; Relaxation Times, τ_j, and Amplitudes, S_j, of Process j; Cole-Cole Shape Parameter, α, of the First (Lower Frequency) Process; Infinite Frequency Permittivity, ε_∞, and Reduced Error Function of the Overall Fit, χ_r^2.

x_{IL}	ε	S_1	τ_1/ps	α	S_2	τ_2/ps	S_3	τ_3/ps	S_4	τ_4/ps	ε_∞	$\chi_r^2/10^{-4}$
0^a	32.49	-	-	-	26.3	52.2	1.27	8.23	1.93	1.05	2.95	
0.007410	31.69	1.86	207	0^b	23.8	50.3	0.783	7.50	2.11	1.69	3.18	12.6
0.01543	30.00	4.34	90.7	0^b	19.6	46.4	1.15	4.96	2.49	1^b	2.44	16.1
0.02435	29.14	3.33	106	0^b	19.5	46.3	1.55	4.63	1.96	1^b	2.79	10.8
0.03355	27.86	4.67	84.5	0^b	16.8	43.9	1.63	4.42	1.94	1^b	2.82	7.46
0.05717	24.71	4.71	61.4	0^b	13.2	43.3	1.63	5.76	1.93	1.42	3.23	22.3
0.08589	22.42	6.55	59.9	0^b	9.04	37.0	2.11	4.26	1.36	1.15	3.37	17.6
0.1242	19.37	11.8	43.6	0^b	0.831	14.0	1.77	4.15	1.90	1^b	3.04	30.6
0.1755	17.44	5.38	51.5	0.03	4.88	34.0	2.14	4.91	1.59	1^b	3.45	6.25
0.2480	14.72	6.19	44.7	0.001	1.45	20.4	1.78	5.00	2.11	1^b	3.19	12.0
0.3624	13.63	1.29	157	0.09	4.71	39.3	1.77	8.15	1.78	1.69	4.08	8.47
0.5583	11.66	4.92	50.2	0.13	-	-	1.27	6.57	1.88	1.12	3.60	6.34
1^c	14.6	10.0	284	0.52	-	-	-	-	2.04	0.620	2.57	

^{a} Parameters from Section 3.3.
^{b} Parameter fixed during fitting procedure.
^{c} Parameters from ref. 38.

in the neat components in the 100 GHz region. Both, τ_1 (Figure 5.18b) and τ_2 decrease with increasing x_{IL}, followed by a subsequent increase for τ_1 until it approaches (with high scatter of the data) the value of the neat IL ($\tau_1 = 284$ ps). The values of τ_2 approach that for neat MeOH (52.2 ps) at low IL content, whereas at $x_{\text{IL}} \gtrsim 0.4$ this process is no longer detectable.

The amplitude of process 1, S_1, increases strongly at low IL content (Figure 5.18a) but scatters considerably at $x_{\text{IL}} \gtrsim 0.3$. Together with the composition dependence of the relaxation times, a superposition of ion pair and cation relaxations is suggested at low x_{IL}. The values of S_2 decrease strongly down to the detection limit of process 2. For process 3, S_3 increases with increasing IL content, passes through a poorly defined maximum at $x_{\text{IL}} \approx 0.2$, and then decreases until $S_3 = 0$. The values for S_4 do not show significant variation with respect to composition.

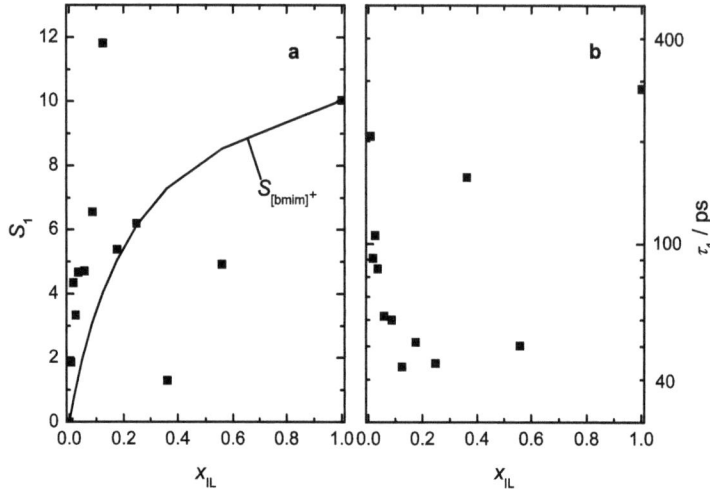

Figure 5.18: (a) Amplitudes of relaxation process 1, S_1, (symbols) and the cation relaxation, $S_{[bmim]^+}$, estimated from Eq. 1.68, (full line), together with (b) the corresponding relaxation time, τ_1, of [bmim][BF$_4$] + MeOH mixtures at 25 °C.

Discussion

MeOH relaxations. In neat MeOH (Figure 3.3), the low-frequency relaxation is associated with the cooperative H-bond relaxation of the oligomeric MeOH chains. In the IL + MeOH mixtures, this relaxation corresponds to process 2. The smooth decrease of τ_2, starting from the value found in neat MeOH, indicates a structure-breaking effect of the added IL. This is consistent with the comparison of the experimental values of the analytical concentration of MeOH, c_{MeOH}, and the observed c_2, the concentration estimated from Eq. 1.68 assuming that all MeOH molecules contribute to the spectrum and that their dipole moment, $\mu_{eff,MeOH}$, has the same value as in neat MeOH (Figure 5.19a). As for the IL + AN mixtures, a strong decrease of c_2 together with $c_2 < c_{MeOH}$ was found in the composition range studied. This is most probably the result of irrotational bonding, i.e. an immobilization on the DR timescale of MeOH molecules. Note, that in the present analysis, no indications for the existence of a 'slow' solvent process have been found. Irrotational bonding is indeed suggested by MD simulations of

5.2. DIELECTRIC PROPERTIES

IL + MeOH mixtures,[33] where strong interactions of MeOH notably with the anions, but also with the charged part of the cation, have been observed. The values of the effective solvation numbers, Z_{ib}, calculated according to Eq. 5.7, show a strong decrease with respect to IL content (Figure 5.19b). Extrapolation to infinite dilution yields the relatively large value of $Z_b^\circ = 13 \pm 1$.

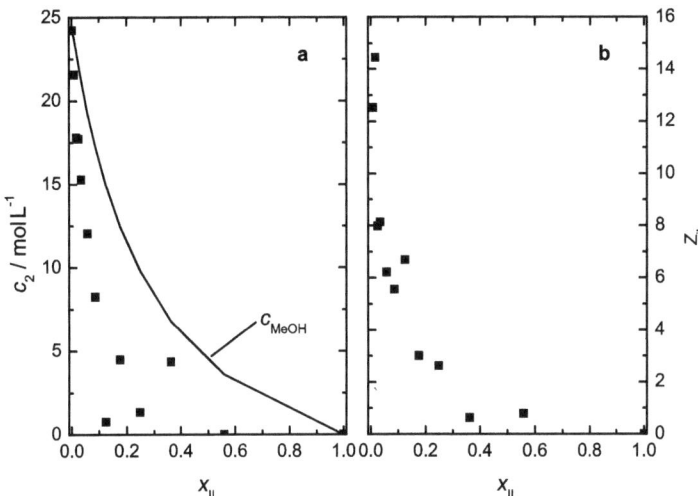

Figure 5.19: **(a)** Molar concentration of MeOH calculated via Eq. 1.68 for process 2, c_2 (symbols), together with the analytical MeOH concentration, c_{MeOH} (line), and **(b)** solvation numbers, Z_{ib}, obtained for binary [bmim][BF$_4$] + MeOH mixtures at 25 °C as a function of the IL mole fraction, x_{IL}.

The medium-frequency relaxation of neat MeOH, with its center located at ~ 20 GHz, has been ascribed to rotational diffusion of individual methanol molecules.[166] It is also present in [bmim][BF$_4$] + MeOH mixtures, here labelled process 3. The lack of data for the viscosities of these mixtures permits a further analysis of τ_3. Nevertheless, S_3 can be related to the concentration of free MeOH molecules by using Eq. 1.68 and assuming that their dipole moment has the same value as in neat MeOH. The results show, that the amount of free MeOH molecules increases almost linearly from $\sim 3\%$ in dilute solutions to $\sim 30\%$ at $x_{IL} = 0.55$. This implies a dramatic shortening of the oligomeric chains in the presence of the IL.

Due to its location at the high-frequency limit of the present spectra, τ_4 occasionally had to be fixed during the fitting procedure. The overlap of several expected modes from the IL as well as from MeOH (Section 3) precludes a further analysis.

IL relaxation. Analogous to IL + AN mixtures, $\mu_{\text{eff},1}$ strongly decreases at low x_{IL}. Together with the composition dependence of S_1 and τ_1 (Figure 5.18), this suggests the coexistence of ion pairs and cations. Ion pairs have indeed been observed by mass spectrometry in [bmim][PF$_6$] + MeOH mixtures.[29] Extrapolation of $\mu_{\text{eff},1}$ to infinite dilution yields the relatively small value of $\mu_{\text{app,CIP}} = 12.2$ D for a [bmim][BF$_4$] ion pair in MeOH. This value can be used in Eq. 1.84 (where $g_j = 1$ and thus $\mu_{\text{eff},j} = \mu_{\text{ap},j}$ is assumed) to estimate the concentrations of ion pairs and the corresponding association constants (Eq. 1.86). Again, the scatter of the values so obtained is high. Nevertheless, the results show a strong decrease of $c_{\text{CIP}}/c_{\text{IL}}$ and K_A with increasing IL content. At $x_{\text{IL}} \gtrsim 0.2$, no ion pairs are observable. By extrapolation with Eq. 1.87, the standard (infinite dilution) association constant for the formation of the CIPs, K_A°, can be estimated. For MeOH at 25 °C, the Debye-Hückel coefficients are $A_{\text{DH}} = 1.914$ L$^{1/2}$ mol$^{-1/2}$ and $B_{\text{DH}} = 5.110 \cdot 10^9$ L$^{1/2}$ mol$^{-1/2}$ m^{-1}. The value so obtained for $\log K_A^\circ = 2.65$ is much higher than the result from conductivity measurements of [bmim][BF$_4$] in MeOH ($\log K_A^\circ = 1.576$).[89] This discrepancy might be a consequence of the unrealistically low value of 12.2 D for $\mu_{\text{app,CIP}}$. For example, performing the analysis with $\mu_{\text{app,CIP}} = 21.1$ D (as for [bmim][BF$_4$] + AN mixtures), one obtains $\log K_A^\circ = 0.940$, which is considerably lower than the conductivity value. Compared to [bmim][BF$_4$] + AN mixtures, the lower value for $\log K_A^\circ$ is more plausible, because firstly, the ion pair concentrations are less in MeOH and secondly, solvation of the ions is expected to be more pronounced in MeOH due to formation of hydrogen bonds.

The overlap of CIP and cation contributions in process 1 does not permit quantitative analysis of the ion pair kinetics for [bmim][BF$_4$] + MeOH mixtures.

5.3 Raman spectroscopy of IL + acetonitrile mixtures

The present Raman spectra for [emim][EtSO$_4$] + AN mixtures over the whole composition range are available in the frequency range $50 \leq \bar{\nu}/\text{cm}^{-1} \leq 4000$. The corresponding measurements for AN and [emim][EtSO$_4$] have been compared to available literature spectra[196,212,213] and general a high level of agreement was found. For the neat components, all bands observed in the experimental spectra have been assigned.[196,213] Among these, the CN-stretch of AN at $\sim$2253 cm^{-1} in the spectra of [emim][EtSO$_4$] + AN mixtures (Figure 5.20a) was found to be most promising for further analysis, as it was sufficiently well separated from neighboring bands and showed a distinct variation with composition. According to ref. 196, the mode at $\sim$2253 cm^{-1} consists of two single bands, the CN stretching mode at 2253.6 cm^{-1} arising from free AN molecules and a hot band attributed to the C − C − N bending mode.[196,197] A combination band arising from the symmetric bend of CH$_3$ and the C−C stretch is located close to the stretching vibration at 2292.7 cm^{-1}.[196,212]

Among the various functions and combinations thereof available to fit spectroscopic data, a combination of three Gauss-Lorentz product functions, the 3 G·L model, described the present spectra. This model corresponds to the formula

$$I(\bar{\nu}) = \sum_{j=1}^{3} \frac{a_j}{1 + s_j \left(\frac{\bar{\nu}-c_j}{w_j}\right)^2 \exp\left[\frac{1}{2}(1-s_j)\left(\frac{\bar{\nu}-c_j}{w_j}\right)^2\right]}, \quad (5.10)$$

where a_j, c_j, w_j and s_j are the amplitude, center, width and shape of the jth band ($j = 1\ldots 3$). The parameters obtained for fitting the Raman spectra in the range $2150 \leq \bar{\nu}/\text{cm}^{-1} \leq 2350$ are listed in Table 5.14. A linear baseline was assumed in the fitting procedure.

The combination band, labelled 3 in Figure 5.20b and Table 5.14, was taken into account in the fitting procedure to achieve stable fits of the bands of interest at $\sim$2253 cm^{-1}. The amplitude of band 3, a_3, decreases linearly with increasing IL concentration and for high x_{IL} it cannot be resolved. This suggests a smooth transition from electrolyte solution to IL-like behavior but, because of its complex nature and low intensity, it will not be further analyzed.

The location of the centers of both bands 1 and 2, c_1 and c_2, vary smoothly and decrease linearly with IL concentration (Figure 5.21a). The respective amplitudes a_1 and a_2 show a strong linear decrease with increasing IL concentration (Figure 5.21b). The same linear decrease was found for the corresponding areas of bands 1 and 2, f_1 and f_2, (not shown) as well as for their sum, $f_1 + f_2$, the total area of the band at $\sim$2253 cm^{-1}.

For Raman spectra of AN + water mixtures[214] and for IR spectra of electrolyte solutions in AN, a blue-shift of the CN stretching band was observed, consequently leading to the formation of new bands.[196,197,215] This has been interpreted in terms of solvation of the cations by the solvent molecules. Contrary to these findings and the present DR results, both, the centers of the CN stretching and the hot bands are shifted to lower frequencies (Figure 5.21a), indicating a weak solvation tendency of the [emim]$^+$ cations by AN. To explain this effect, one could imagine a reduction of the CN force constants (red-shift) for AN molecules interacting with nonpolar IL domains[135] via their methyl group. Assuming the magnitude of this red-shift to be more pronounced than the blue-shift expected for solvating AN molecules, the overall band shift would result in an decrease of the CN stretching band centers. The linear variation of the band centers, amplitudes (Figure 5.21) and areas indicate a gradual change of the liquid structure of the present mixtures. In other words, the present analysis implies a more or less simple dilution of the AN by the IL. However, more studies will be needed to shed more light onto the vibrational nature of IL + polar solvent mixtures.

5.3. RAMAN SPECTROSCOPY OF IL + ACETONITRILE MIXTURES

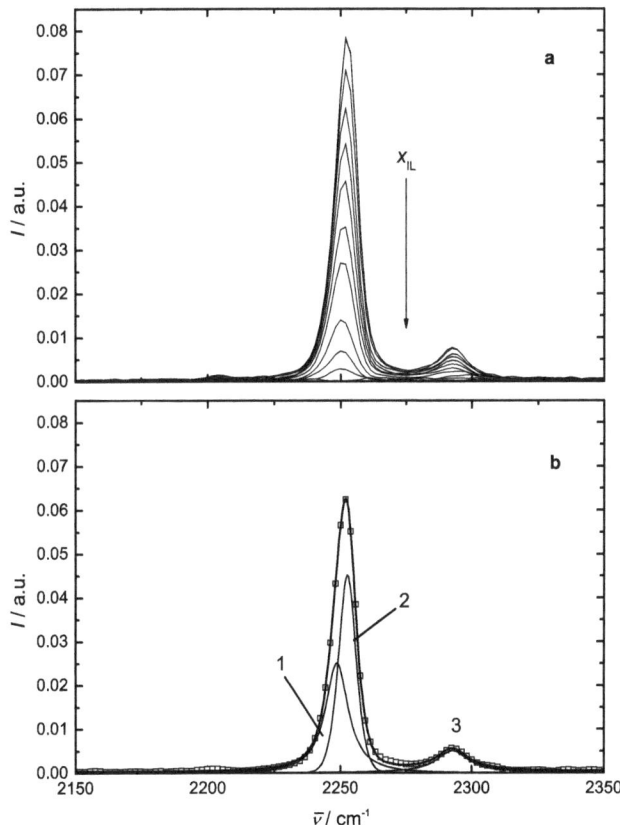

Figure 5.20: Experimental Raman spectra, $I(\bar{\nu})$, of [emim][EtSO$_4$] + AN mixtures. Lines in (a) and symbols in (b) represent experimental data, lines in (b) show the overall fit with the 3 G·L model and the contributions of the individual processes for a selected mixture ($x_{\mathrm{IL}} = 0.1036$).

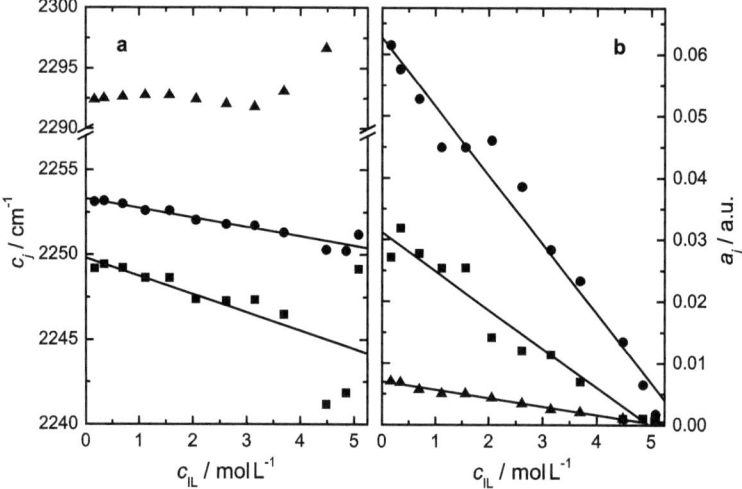

Figure 5.21: (a) Band centers, c_j, and (b) amplitudes, a_j, of band 1 (■), 2 (●) and 3 (▲) obtained from fitting the 3 G·L model to Raman spectra of [emim][EtSO$_4$] + AN mixtures. Lines are linear fits.

5.3. RAMAN SPECTROSCOPY OF IL + ACETONITRILE MIXTURES

Table 5.14: Fit Parameters of Eq. 5.10 for the Observed Raman Spectra of Mixtures of [emim][EtSO4] with AN Assuming the 3 G-L Model: Amplitudes, a_j, Centers, c_j, Widths, w_j, Shapes, s_j, and Areas, f_j, of Band j; Standard Deviation of the Overall Fit, σ.[a] Note, that j here refers to the specified Raman bands as defined the the text.

x_{IL}	a_1 a.u.	c_1 cm^{-1}	w_1 cm^{-1}	s_1	f_1 a.u.	a_2 a.u.	c_2 cm^{-1}	w_2 cm^{-1}	s_2	f_2 a.u.	a_3 a.u.	c_3 cm^{-1}	w_3 cm^{-1}	s_3	f_3 a.u.	$10^4 \cdot \sigma$ a.u.
0.008940	0.027	2249.2	5.1	0.99	0.40	0.061	2253.1	3.9	0.63	0.56	0.0071	2292.4	6.5	0.98	0.13	3
0.01904	0.032	2249.4	5.0	0.99	0.47	0.058	2253.2	3.8	0.63	0.51	0.0068	2292.5	6.8	1.0	0.14	3
0.04008	0.028	2249.2	5.0	1[b]	0.42	0.053	2253.0	3.9	0.64	0.48	0.0057	2292.6	7.1	1[b]	0.12	4
0.06873	0.025	2248.6	5.2	0.97	0.37	0.045	2252.6	3.8	0.51	0.40	0.0051	2292.8	7.5	1.0	0.11	3
0.1036	0.025	2248.6	5.2	0.97	0.37	0.045	2252.6	3.8	0.51	0.40	0.0051	2292.8	7.5	1.0	0.11	3
0.1482	0.014	2247.4	5.3	1[b]	0.23	0.046	2252.1	4.2	0.54	0.45	0.0044	2292.4	6.8	1[b]	0.089	3
0.2100	0.012	2247.3	5.4	1[b]	0.20	0.039	2251.8	4.2	0.49	0.38	0.0035	2292.0	8.2	1[b]	0.083	2
0.2861	0.011	2247.3	5.5	1[b]	0.19	0.028	2251.7	4.2	0.48	0.28	0.0025	2291.8	10	1[b]	0.073	2
0.3866	0.0070	2246.5	5.4	1[b]	0.11	0.023	2251.3	4.4	0.43	0.24	0.0021	2293.1	8.9	0.92	0.048	2
0.6063	0.0010	2241.2	3.9	1[b]	0.012	0.013	2250.3	4.9	0.29	0.15	0.0011	2296.6	5.5	1[b]	0.017	2
0.7645	0.0011	2241.9	2.9	1[b]	0.0095	0.0065	2250.2	4.6	0.30	0.070	–	–	–	–	–	2
0.8924	9.0E-4	2249.2	3.2	1[b]	0.0088	0.0018	2251.2	5.7	0.20	0.024	–	–	–	–	–	1

[a] Areas in arbitrary units (a.u.); widths are full band widths at half maximum peak height; shapes are Gauss-Lorentz mixing fractions.
[b] Parameter fixed during fitting procedure.

Summary and conclusions

Neat components

The present thesis has highlighted recent insights gained largely by DR spectroscopy into the structure and dynamics of neat ionic liquids (ILs), because their knowledge was essential for the analysis of DR spectra of IL + IL and IL + solvent mixtures. Most importantly, the presence of mesoscale structure and inhomogenity in ILs was experimentally confirmed by comparison of dielectric relaxation (DR) and optical heterodyne-detected Raman-induced Kerr effect spectra covering a large frequency range.[39] The nature of the cation reorientation through large-angle jumps is now understood. Of special relevance for the present PhD thesis, Hunger[40] could show from analysis of broadband DR spectra that microwave modes are still well characterized when spectra were limited to $\nu \leq 89$ GHz.

Application of THz spectroscopy and subsequent combination of the spectrum obtained with various published data yielded an exceptionally wide frequency coverage from 0.1 GHz to ~ 10 THz for acetonitrile (AN) at 25 °C. As an important result of the temperature dependent DR study performed in a more limited frequency range ($\nu \leq 89$ GHz), the main relaxation step was related to the rotational diffusion of AN dipoles under slip boundary conditions as the mechanism for dielectric relaxation. Of special relevance for the present studies it was shown, that the main relaxation in the microwave region is still reliably characterized even though spectra are limited to ≤ 89 GHz. For the two THz modes, various possible molecular-level motions were discussed, but most probably, intermolecular vibrations and librations mainly contribute at high frequencies.

The DR spectrum of methanol (MeOH) at GHz to THz frequencies was published and analyzed in terms of dynamic features by Fukasawa *et al.*[166] To allow quantitative analysis of IL + MeOH mixtures, the published spectrum was fit applying the model given by the authors and the underlying molecular-level motions were briefly reviewed.

IL + IL mixtures

Dielectric relaxation spectroscopy has for the first time been used to investigate the structure and dynamics of mixtures of two ILs sharing a common cation. The DR spectra of [emim][BF$_4$] + [emim][DCA] mixtures in the frequency range $0.2 \leq \nu/\text{GHz} \leq 20$ were well fit by only two processes. The lower-frequency relaxation is mostly associated with cation jump reorientation, and the Debye relaxation at $\sim 100\,\text{GHz}$ represents the low-frequency wing of much higher frequency intermolecular vibrations and librations.

The negligible excess volumes, as well as viscosities, effective dipole moments and CC broadness parameters, show very close to 'ideal' behavior. That is, η, $\mu_{\text{eff},+}$ and α show close to linear variation with composition, with only small 'excess' quantities. This suggests a gradual change of the liquid structure between the end members. On the other hand, enhanced rotational (τ_1) and translational (κ) dynamics were observed for these mixtures.

The values of the effective dipole moments of the neat components suggested strong antiparallel dipole-dipole correlations among the cations especially for the dicyanamide salt. The present data for [emim][BF$_4$] are compatible with simulations of de Andrade et al.[179] indicating pronounced cation stacking. It is speculated that the structure of [emim][DCA] might be dominated by cation dimers capped by anions, similar to that proposed for [emim][AlCl$_4$].[179] Computer simulations using the present data would be of interest.

IL + polar solvent mixtures

The present PhD thesis has presented reliable values for the densities, conductivities, molar conductivities, and excess molar volumes of selected IL + AN and IL + MeOH mixtures, which have established that much of the available literature data are of doubtful quality. Of major importance, the purity of ILs is a crucial topic. Quite often, the purity of the samples used in published studies was not properly determined or specified. Just in recent time, much more efforts are made, particularly by commercial suppliers, to overcome this unsatisfactory situation.

The present DR study, covering the frequency range of $0.2 \lesssim \nu/\text{GHz} \leq 89$, shed more light into the structure, speciation and dynamics of binary mixtures of ILs with AN or MeOH. The spectra can be fit over the entire composition range by consistent models. However, detailed analysis showed that these processes are superpositions of various modes.

The lowest-frequency process associated with large-angle jump reorientation of the cations is overlapped at $x_{\text{IL}} \lesssim 0.3$ by a contribution from the reorientation of contact ion pairs and, for IL + AN mixtures, by AN molecules, that are slowed (cf. neat AN) on the DR timescale. For

MeOH, considerable irrotational bonding was observed, probably as a consequence of strong H-bond interactions with the anions. The picture that emerged from the analysis is consistent with recent computer simulations.[32,33,135] They proposed microphase segregation in ILs and balanced interactions of AN and MeOH with the nonpolar and polar domains present in ILs. Both, AN and MeOH were found to strongly interact with the ions, with the interaction being less directional for AN. These predictions are strongly supported by our data.

Among all systems studied, [emim][EtSO$_4$] + AN mixtures differ somewhat from other IL + AN mixtures. An additional medium-frequency process was observed, which could be attributed to a superposition of reorientations of the highly dipolar anions and 'slow' AN molecules. Due to strong mode overlap, the scatter of the derived fit parameters is high. Hence, a quantitative analysis of the anion mode is not possible, but comparison with neat [emim][EtSO$_4$] suggests strong dipole correlation, yielding anion correlation factors $g_- \ll 1$. For the cation, g_+ could be quantitatively determined, showing parallel alignment of the dipoles.

As found previously for IL + dichloromethane mixtures,[37,48] the ILs retain their molten-salt like character, at least in terms of dielectric properties, up to high dilution with AN ($x_{\text{IL}} \gtrsim 0.5$). For MeOH, strong spectral overlap limited a detailed analysis at high IL content, but a similar behavior was suggested within the high scatter of the fit parameters. At high dilution, the ILs behave as conventional, moderately associated electrolytes with significant amounts of contact ion pairs being formed. There were no indications for the existence of solvent-separated ion pairs or higher aggregates in the present analysis. Note, that for neat ILs, the existence of stable ion pairs with a life-time longer than some hundreds of picoseconds has been excluded.[46] In essence, IL + AN or + MeOH mixtures can be divided into two regions. At low IL concentrations they behave as conventional electrolyte solutions with moderately solvated ions and ion pairs, while at higher concentrations they keep the IL-like character. The transition region ($0.3 \lesssim x_{\text{IL}} \lesssim 0.5$) is characterized by redissociation of the ion pairs and establishment of an IL-like structure.

The present analysis of the CN stretching band of the Raman spectra of [emim][EtSO$_4$] + AN mixtures showed a gradual change of the liquid structure, which may be interpreted as smooth transition from molten-salt like to electrolyte solution behavior. The red-shift of the CN stretching band is uncommon compared with conventional electrolyte solutions in AN, but a possible explanation assuming interactions of the AN methyl groups with nonpolar IL domains, as predicted by MD simulations,[135] was given.

Appendix

A.1 Physico-chemical data for [emim][EtSO$_4$] + acetonitrile mixtures

The following tables list densities, viscosities and refractive indices obtained for selected [emim]-[EtSO$_4$] + AN mixtures. Measurements were performed with a DMA 5000 M density meter, an AMVn automated micro falling ball viscometer and a modified Abbemat WR MW automatic digital refractometer (Anton Paar GmbH, Graz, Austria). For the viscosity measurements, the mean value of two measurements at different angles (50° and 60°) together with $\delta\eta = \eta_{50°} - \eta_{60°}$ are given. Details of the experimental setup are given in Section 2.3. Some of the data have been used in the analysis of the DR spectra of [emim][EtSO$_4$] + AN mixtures.

We appreciate the valuable support by Prof. Dr. Augustinus Asenbaum and Dr. Christian Pruner, Universität Salzburg, as well as by C. Schöggl-Wagner and T. Feischl, Anton Paar Gmbh, Graz.

Table A.1: Investigated Mole Fractions, x_{IL}, Densities, ρ, Viscosities, η, and Differences of Two Measurements of η, $\delta\eta$, of Binary [emim][EtSO$_4$] + AN Mixtures at 25 °C.

x_{IL}	ρ / g L^{-1}	η / mPa s	$\delta\eta$ / mPa s
0	776.71	0.3326	0.0004
0.04004	846.68	0.4970	0.0055
0.1036	926.55	0.9370	0.015
0.2862	1066.29	3.799	0.063
0.6059	1177.11	23.18	0.040
0.7637	1206.97	44.36	0.23
0.8913	1224.95	69.78	0.29
1	1238.12	94.62	0.14

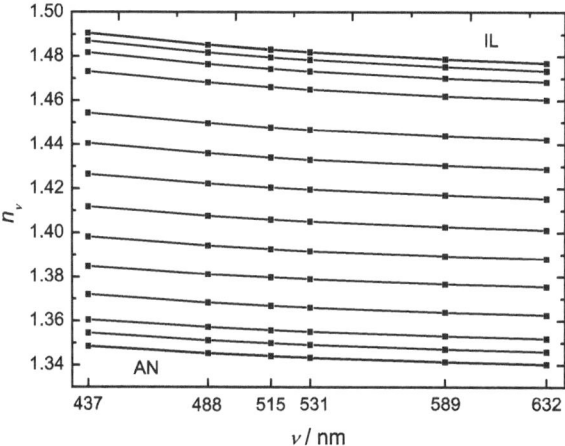

Figure A.1: Dispersion of the refractive indices, n_ν, for [emim][EtSO$_4$] + AN mixtures (Table A.2; increasing IL content from bottom to top) at 25 °C.

Table A.2: Investigated Mole Fractions, x_{IL}, and Refractive Indices, n_ν, Measured at ν/nm = 437, 488, 515, 531, 589 and 632 for Binary [emim][EtSO$_4$] + AN Mixtures at 25 °C.

x_{IL}	n_{437}	n_{488}	n_{515}	n_{531}	n_{589}	n_{632}
0	1.34848	1.34528	1.34399	1.34328	1.34135	1.34022
0.008951	1.35447	1.35121	1.34987	1.34913	1.34707	1.34592
0.01903	1.36042	1.35713	1.35579	1.35506	1.35306	1.35188
0.04006	1.37183	1.36827	1.36680	1.36598	1.36383	1.36253
0.06874	1.38476	1.38115	1.37966	1.37887	1.37665	1.37539
0.1035	1.39808	1.39415	1.39253	1.39166	1.38933	1.38807
0.1484	1.41175	1.40763	1.40594	1.40503	1.40256	1.40115
0.2098	1.42655	1.42224	1.42049	1.41956	1.41697	1.41541
0.2861	1.44055	1.43596	1.43413	1.43312	1.43037	1.42879
0.3872	1.45440	1.44965	1.44774	1.44672	1.44385	1.44219
0.6053	1.47314	1.46816	1.46612	1.46500	1.46196	1.46021
0.7643	1.48170	1.47646	1.47439	1.47323	1.47010	1.46836
0.8912	1.48693	1.48163	1.47954	1.47838	1.47521	1.47340
1	1.49060	1.48523	1.48312	1.48195	1.47875	1.47692

Table A.3: Measured Refractive Indices, n_{589}, at $\nu = 589$ nm for Neat [emim][EtSO$_4$] as Function of Temperature, θ.

$\theta/°\mathrm{C}$	n_{589}
10	1.48271
15	1.48138
25	1.47875
35	1.47605
45	1.47338
55	1.47079
65	1.46810

Bibliography

[1] Wasserscheid, P.; Welton, T. *Ionic Liquids in Synthesis*; Wiley-VCH: Weinheim, Germany, 2003.

[2] Zistler, M.; Wachter, P.; Wasserscheid, P.; Gerhard, D.; Hinsch, A.; Sastrawan, R.; Gores, H. J. *Electrochim. Acta* **2006**, *52*, 161.

[3] Wachter, P.; Zistler, M.; Schreiner, C.; Fleischmann, M.; Gerhard, D.; Wasserscheid, P.; Barthel, J.; Gores, H. J. *J. Chem. Eng. Data* **2009**, *54*, 491.

[4] Canongia Lopes, J. N.; Cordeiro, T. C.; Esperança, J. M. S. S.; Guedes, H. J. R.; Huq, S.; Rebelo, L. P. N.; Seddon, K. R. *J. Phys. Chem. B* **2005**, *109*, 3519.

[5] Navia, P.; Troncoso, J.; Romaní, L. *J. Chem. Eng. Data* **2007**, *52*, 1369.

[6] Navia, P.; Troncoso, J.; Romaní, L. *J. Chem. Eng. Data* **2007**, *52*, 2542.

[7] Navia, P.; Troncoso, J.; Romaní, L. *J. Solution Chem.* **2008**, *37*, 677.

[8] Fletcher, K. A.; Baker, S. N.; Baker, G. A.; Pandey, S. *New J. Chem.* **2003**, *27*, 1706.

[9] Xiao, D.; Rajian, J. R.; Li, S.; Bartsch, R. A.; Quitevis, E. L. *J. Phys. Chem. B* **2006**, *110*, 16174.

[10] Xiao, D.; Rajian, J. R.; Hines, L. G.; Li, S.; Bartsch, R. A.; Quitevis, E. L. *J. Phys. Chem. B* **2008**, *112*, 13316.

[11] Welton, T. *Chem. Rev.* **1999**, *99*, 2071.

[12] MacFarlane, D. R.; Seddon, K. R. *Aust. J. Chem.* **2007**, *60*, 3.

[13] Martins, M. A. P.; Frizzo, C. P.; Moreira, D. N.; Zanatta, N.; Bonacorso, H. G. *Chem. Rev.* **2008**, *108*, 2015.

[14] Roosen, C.; Müller, P.; Greiner, L. *Appl. Microbiol. Biotechnol.* **2008**, *81*, 607.

[15] Weingärtner, H. *Angew. Chem. Int. Ed.* **2008**, *47*, 654.

[16] Dreyer, S.; Kragl, U. *Biotechnol. Bioeng.* **2008**, *99*, 1416.

[17] Kimizuka, N.; Nakashima, T. *Langmuir* **2001**, *17*, 6759.

[18] Roth, M. *J. Chromatogr. A* **2009**, *1216*, 1861.

[19] Seddon, K. R.; Stark, A.; Torres, M.-J. *Pure Appl. Chem.* **2000**, *72*, 2275.

[20] Marsh, K. N.; Boxall, J. A.; Lichtenthaler, R. *Fluid Phase Equilib.* **2004**, *219*, 93.

[21] Heintz, A. *J. Chem. Thermodyn.* **2005**, *37*, 525.

[22] Stoppa, A.; Hunger, J.; Buchner, R. *J. Chem. Eng. Data* **2009**, *54*, 472.

[23] Chakrabarty, D.; Chakraborty, A.; Seth, D.; Sarkar, N. *J. Phys. Chem. A* **2005**, *109*, 1764.

[24] Mellein, B. R.; Aki, S. N. V. K.; Ladewski, R. L.; Brennecke, J. F. *J. Phys. Chem. B* **2007**, *111*, 131.

[25] Katoh, R.; Hara, M.; Tsuzuki, S. *J. Phys. Chem. B* **2008**, *112*, 15426.

[26] Umebayashi, Y.; Jiang, J.-C.; Lin, K.-H.; Shan, Y.-L.; Fujii, K.; Seki, S.; Ishiguro, S.-I.; Lin, S. H.; Chang, H.-C. *J. Chem. Phys.* **2009**, *131*, 234502.

[27] Consorti, C. S.; Suarez, P. A. Z.; de Souza, R. F.; Burrow, R. A.; Farrar, D. H.; Lough, A. J.; Loh, W.; da Silva, L. H. M.; Dupont, J. *J. Phys. Chem. B* **2005**, *109*, 4341.

[28] Takamuku, T.; Honda, Y.; Fujii, K.; Kittaka, S. *Anal. Sci.* **2008**, *24*, 1285.

[29] Kanakubo, M.; Hiejima, Y.; Aizawa, T.; Kurata, Y.; Wakisaka, A. *Chem. Lett.* **2005**, *34*, 706.

[30] Aliotta, F.; Ponterio, R. C.; Saija, F.; Salvato, G.; Triolo, A. *J. Phys. Chem. B* **2007**, *111*, 10202.

[31] Wu, X.; Liu, Z.; Huang, S.; Wang, W. *Phys. Chem. Chem. Phys.* **2005**, *7*, 2771.

[32] Canongia Lopes, J. N.; Costa Gomes, M. F.; Pádua, A. A. H. *J. Phys. Chem. B* **2006**, *110*, 16816.

[33] Pádua, A. A. H.; Costa Gomes, M. F.; Canongia Lopes, J. N. A. *Acc. Chem. Res.* **2007**, *40*, 1087.

[34] Buchner, R.; Hefter, G. *Phys. Chem. Chem. Phys.* **2009**, *11*, 8954.

[35] Schrödle, S.; Annat, G.; MacFarlane, D. R.; Forsyth, M.; Buchner, R.; Hefter, G. *Chem. Commun.* **2006**, 1748.

[36] Stoppa, A.; Hunger, J.; Thoman, A.; Helm, H.; Hefter, G.; Buchner, R. *J. Phys. Chem. B* **2008**, *112*, 4854.

[37] Hunger, J.; Stoppa, A.; Buchner, R.; Hefter, G. *J. Phys. Chem. B* **2009**, *113*, 9527.

[38] Hunger, J.; Stoppa, A.; Schrödle, S.; Hefter, G.; Buchner, R. *ChemPhysChem* **2009**, *10*, 723.

[39] Turton, D. A.; Hunger, J.; Stoppa, A.; Hefter, G.; Thoman, A.; Walther, M.; Buchner, R.; Wynne, K. *J. Am. Chem. Soc.* **2009**, *131*, 11140.

[40] Hunger, J. Ph.D. thesis, Regensburg, 2010.

[41] Hefter, G.; Buchner, R.; Hunger, J.; Stoppa, A. Chemical Speciation in Ionic Liquids and their Mixtures with Polar Solvents Using Dielectric Spectroscopy. In *Ionic Liquids: From Knowledge to Applications*; Rogers, R. D., Plechkova, N. V., Seddon, K. R., Eds.; ACS Symposium Series: Washington, 2010; Vol. 1030, pp 61–74.

[42] *Broadband Dielectric Spectroscopy*; Kremer, F., Schönhals, A., Eds.; Springer: Berlin, 2003.

[43] Marcus, Y.; Hefter, G. *Chem. Rev.* **2006**, *106*, 4585.

[44] Buchner, R. *Pure Appl. Chem.* **2008**, *80*, 1239.

[45] Fraser, K. J.; Izgorodina, E. I.; Forsyth, M.; Scott, J. L.; MacFarlane, D. R. *Chem. Commun.* **2007**, 3817.

[46] Sangoro, J.; Iacob, C.; Serghei, A.; Naumov, S.; Galvosas, P.; Kärger, J.; Wespe, C.; Bordusa, F.; Stoppa, A.; Hunger, J.; Buchner, R.; Kremer, F. *J. Chem. Phys.* **2008**, *128*, 214509.

[47] Zhao, W.; Leroy, F.; Heggen, B.; Zahn, S.; Kirchner, B.; Balasubramanian, S.; Müller-Plathe, F. *J. Am. Chem. Soc.* **2009**, *131*, 15825.

[48] Hunger, J.; Stoppa, A.; Hefter, G.; Buchner, R. *J. Phys. Chem. B* **2008**, *112*, 12913.

[49] Hunger, J.; Zahn, S.; Uhlig, F.; Stoppa, A.; Buchner, R.; Kirchner, B. unpublished results.

[50] Smith, J. W. *Electric Dipole Moments*; Butterworth Scientific Publications: London, UK, 1955.

[51] Parker, A. J. *Pure Appl. Chem.* **1981**, *53*, 1437.

[52] Chu, A.; Braatz, P. *J. Power Sources* **2002**, *112*, 236.

[53] Snyder, L. R.; Kirkland, J. J.; Dolan, J. W. *Introduction to Modern Liquid Chromatography*, 3rd ed.; Wiley: Hoboken, USA, 2010.

[54] Barthel, J.; Neueder, R.; Schröder, P. In *Electrolyte Data Collection, Part 1c: Conductivities, Transference Numbers, Limiting Ionic Conductivities of Solutions of Aprotic, Protophobic Solvents. I: Nitriles*; Kreysa, G., Ed.; Dechema: Frankfurt, 1996; Vol. XII.

[55] Marcus, Y. *Ion Properties*; Marcel Dekker: New York, USA, 1997.

[56] Tubbs, J. D.; Hoffmann, M. M. *J. Solution Chem.* **2004**, *33*, 381.

[57] Barthel, J.; Neueder, R. In *Electrolyte Data Collection, Part 1: Conductivities, Transference Numbers, Limiting Ionic Conductivities.*; Eckermann, R., Kreysa, G., Eds.; Dechema: Frankfurt, 1992; Vol. XII.

[58] Maxwell, J. C. *Treatise in Electricity and Magnetism*; Claredon Tress: Oxford, 1881.

[59] Greschner, G. S. *Maxwellgleichungen*; Hüthig: Basel, 1981.

[60] Böttcher, C. F. J.; Bordewijk, P. *Theory of Electric Polarization*; Elsevier: Amsterdam, 1978; Vol. 1 and 2.

[61] Barthel, J.; Buchner, R.; Steger, H. *Wiss. Zeitschr. THLM* **1989**, *31*, 409.

[62] Debye, P. *Polar Molecules*; Dover Publ.: New York, 1930.

[63] Pellat, H. *Ann. Chim. Phys.* **1899**, *18*, 150.

[64] Cole, K. S.; Cole, R. H. *J. Chem. Phys.* **1941**, *9*, 341.

[65] Cole, K. S.; Cole, R. H. *J. Chem. Phys.* **1942**, *9*, 98.

[66] Turton, D. A.; Wynne, K. *J. Chem. Phys.* **2008**, *128*, 154516.

[67] Davidson, D. W.; Cole, R. H. *J. Chem. Phys.* **1950**, *18*, 1417.

[68] Davidson, D. W.; Cole, R. H. *J. Chem. Phys.* **1951**, *19*, 1484.

[69] Havriliak, S.; Negami, S. *J. Polym. Sci., Part C* **1966**, *14*, 99.

[70] Chalmers, J. M.; Griffiths, P. R. *Handbook of Vibrational Spectroscopy*; Wiley–VCH: Weinheim, 2001.

[71] Schrödle, S. Ph.D. thesis, Regensburg, 2005.

[72] Bevington, P. R. *Data Reduction and Error Analysis for the Physical Sciencies*; Mc-Graw–Hill: New York, 1969.

[73] Steger, H. Ph.D. thesis, Regensburg, 1988.

[74] Onsager, L. *J. Am. Chem. Soc.* **1936**, *58*, 1486.

[75] Kirkwood, J. G. *J. Chem. Phys.* **1939**, *7*, 911.

[76] Fröhlich, H. *Theory of Dielectrics*, 2nd ed.; Oxford University Press: Oxford, 1965.

[77] Cavell, E. A. S.; Knight, P. C.; Sheikh, M. A. *Trans. Faraday Soc.* **1971**, *67*, 2225.

[78] Barthel, J.; Buchner, R. *Chem. Soc. Rev.* **1992**, *21*, 263.

[79] Scholte, T. G. *Physica* **1949**, *15*, 437.

[80] Barthel, J.; Hetzenauer, H.; Buchner, R. *Ber. Bunsen–Ges. Phys. Chem.* **1992**, *96*, 1424.

[81] Kivelson, D.; Madden, P. *Annu. Rev. Phys. Chem.* **1980**, *31*, 523.

[82] Dote, J. C.; Kivelson, D.; Schwartz, R. N. *J. Phys. Chem.* **1981**, *85*, 2169.

[83] Perrin, F. *J. Phys. Radium* **1934**, *5*, 497.

[84] Dote, J. C.; Kivelson, D. *J. Phys. Chem.* **1983**, *87*, 3889.

[85] Barthel, J.; Kleebauer, M.; Buchner, R. *J. Solution Chem.* **1995**, *24*, 1.

[86] Powles, J. G. *J. Chem. Phys.* **1953**, *21*, 633.

[87] Glarum, S. H. *J. Chem. Phys.* **1960**, *33*, 639.

[88] Madden, P.; Kivelson, D. *Adv. Chem. Phys.* **1984**, *56*, 467.

[89] Bešter-Rogač, M.; Hunger, J.; Stoppa, A.; Buchner, R. *J. Chem. Eng. Data* **2010**, *55*, 1799.

[90] Buchner, R.; Barthel, J. *J. Mol. Liq.* **1995**, *63*, 55.

[91] Strelow, H.; Knoche, W. *Fundamentals of Chemical Relaxations*; Verlag Chemie: Weinheim, 1977.

[92] Cussler, E. L. *Diffusion: Mass Transfer in Fluid Systems*; Cambridge University Press: Cambridge, 1986.

[93] Moore, W. J.; Hummel, D. O. *Physikalische Chemie*; de Gruyter: Berlin, 1986.

[94] Glasstone, S.; Laidler, K. J.; Eyring, H. *The Theory of Rate Processes*; McGraw Hill: New York, 1977.

[95] Doolittle, A. K. *J. Appl. Phys.* **1951**, *87*, 1471.

[96] Angell, C. A. *Science* **1995**, *267*, 1924.

[97] Stoppa, A.; Zech, O.; Kunz, W.; Buchner, R. *J. Chem. Eng. Data* **2010**, *55*, 1768.

[98] Hünig, S.; Kreitmeier, P.; Märkl, G.; Sauer, J. *Arbeitsmethoden in der organischen Chemie*; Verlag Lehmanns: Berlin, 2006.

[99] Schleicher, J. C.; Scurto, A. M. *Green Chem.* **2009**, *11*, 694.

[100] Holbrey, J. D.; Seddon, K. R. *J. Chem. Soc. Dalton Trans.* **1999**, *13*, 2133.

[101] Lancaster, N. L.; Welton, T.; Young, G. B. *J. Chem. Soc. Perkin Trans.* **2001**, *2*, 2267.

[102] Cammarata, L.; Kazarian, S. G.; Slater, P. A.; Welton, T. *Phys. Chem. Chem. Phys.* **2001**, *3*, 5192.

[103] Fredlake, C. P.; Crosthwaite, J. M.; Hert, D. G.; Aki, S. N.; Brennecke, J. F. *J. Chem. Eng. Data* **2004**, *49*, 954.

[104] Buchner, R.; Hefter, G.; May, P. M. *J. Phys. Chem. A* **1999**, *103*, 1.

[105] Levine, H.; Papas, C. H. *J. Appl. Phys.* **1951**, *22*, 29.

[106] Blackham, D. V. *IEEE Trans. Instr. Meas.* **1997**, *46*, 1093.

[107] Wölbl, J. Ph.D. thesis, Regensburg, 1982.

[108] Barthel, J.; Buchner, R.; Wurm, B. *J. Mol. Liq.* **2002**, *98-99*, 51.

[109] Stoppa, A. Diploma thesis, Regensburg, 2006.

[110] Schrödle, S.; Hefter, G.; Kunz, W.; Buchner, R. *Langmuir* **2006**, *22*, 924.

[111] Hölzl, C. Ph.D. thesis, Regensburg, 1998.

[112] Barthel, J.; Buchner, R.; Hölzl, C. G.; Münsterer, M. *Z. Phys. Chem.* **2000**, *214*, 1213.

[113] Pickl, H. Ph.D. thesis, Regensburg, 1998.

[114] Barthel, J.; Bachhuber, K.; Buchner, R.; Hetzenauer, H.; Kleebauer, M. *Ber. Bunsen-Ges. Phys. Chem.* **1991**, *95*, 853.

[115] Fischer, B. M. Ph.D. thesis, Freiburg, 2005.

[116] Jepsen, M. U.; Fischer, B. M.; Thoman, A.; Helm, H.; Suh, J. Y.; Lopez, R.; Haglund, R. F. *Phys. Rev. B* **2006**, *74*, 205103.

[117] Thoman, A. Ph.D. thesis, Freiburg, 2009.

[118] Birch, J. R.; Parker, T. J. In *Infrared and Millimeter Waves*; Button, K. J., Ed.; Academic Press: New York, 1979; Vol. 2.

[119] Kratky, O.; Leopold, H.; Stabinger, H. *Z. Angew. Phys.* **1969**, *27*, 273.

[120] *CRC Handbook of Chemistry and Physics*, 85th ed.; Lide, D. R., Ed.; CRC Press: Boca Raton, USA, 2004.

[121] Barthel, J.; Wachter, R.; Gores, H. J. Temperature dependence of conductance of electrolytes in nonaqueous solutions. In *Modern Aspects of Electrochemistry*; Conway, B. E., Bockris, J. O., Eds.; Plenum: New York, 1979; Vol. 13, pp 1–79.

[122] Barthel, J.; Graml, H.; Neueder, R.; Turq, P.; Bernard, O. *Curr. Top. Solution Chem.* **1994**, *1*, 223.

[123] Barthel, J. *Pure Appl. Chem.* **1979**, *51*, 2093.

[124] Barthel, J.; Feuerlein, F.; Neueder, R.; Wachter, R. *J. Solution Chem.* **1980**, *9*, 209.

[125] Hoover, T. B. *J. Phys. Chem.* **1964**, *68*, 876.

[126] Nishida, T.; Tashiro, Y.; Yamamoto, M. *J. Fluorine Chem.* **2003**, *120*, 135.

[127] Zhang, S.; Li, X.; Chen, H.; Wang, J.; Zhang, J.; Zhang, M. *J. Chem. Eng. Data* **2004**, *49*, 760.

[128] Van Valkenburg, M. E.; Vaughn, R. L.; Williams, M.; Wilkes, J. S. *Thermochim. Acta* **2005**, *425*, 181.

[129] Halder, M.; Sanders Headley, L.; Mukherjee, P.; Song, X.; Petrich, J. W. *J. Phys. Chem. A* **2006**, *110*, 8623.

[130] Arzhantsev, S.; Jin, H.; Baker, G. A.; Maroncelli, M. *J. Phys. Chem. B* **2007**, *111*, 4978.

[131] Kobrak, M. N. *Adv. Chem. Phys.* **2008**, *139*, 85.

[132] Horng, M. L.; Gardecki, J. A.; Papazyan, A.; Maroncelli, M. *J. Chem. Phys.* **1995**, *99*, 17311.

[133] Kashyap, H. K.; Biswas, R. *J. Phys. Chem. B* **2008**, *112*, 12431.

[134] Kashyap, H. K.; Biswas, R. *J. Phys. Chem. B* **2010**, *114*, 254.

[135] Canongia Lopes, J. N. A.; Pádua, A. A. H. *J. Phys. Chem. B* **2006**, *110*, 3330.

[136] Wang, Y.; Voth, G. A. *J. Phys. Chem. B* **2006**, *110*, 18601.

[137] Xiao, D.; Rajian, J. R.; Cady, A.; Li, S.; Bartsch, R. A.; Quitevis, E. L. *J. Phys. Chem. B* **2007**, *111*, 4669.

[138] Triolo, A.; Russina, O.; Bleif, H. J.; Di Cola, E. *J. Phys. Chem. B* **2007**, *111*, 4641.

[139] Hu, Z.; Margulis, C. J. *Proc. Natl. Acad. Sci. USA* **2006**, *103*, 831.

[140] Iwata, K.; Okajima, H.; Saha, S.; Hamaguchi, H. *Acc. Chem. Res.* **2007**, *40*, 1174.

[141] Atkin, R.; Warr, G. G. *J. Phys. Chem. B* **2008**, *112*, 4164.

[142] Ohba, T.; Ikawa, S. *Mol. Phys.* **1991**, *73*, 985.

[143] Eberspächer, P. N. Ph.D. thesis, Regensburg, 1996.

[144] Steinhauser, O.; Bertagnolli, H. *Chem. Phys. Lett.* **1980**, *78*, 555.

[145] La Manna, G.; Notaro, C. E. *J. Mol. Liq.* **1992**, *54*, 125.

[146] Radnai, T.; Jedlovszky, P. *J. Phys. Chem.* **1994**, *98*, 5994.

[147] Richardi, J.; Fries, P. H.; Fischer, R.; Rast, S.; Krienke, H. *J. Mol. Liq.* **1997**, *73-74*, 465.

[148] Katayama, M.; Komori, K.; Ozutsumi, K.; Ohtaki, H. *Z. Phys. Chem.* **2004**, *218*, 659.

[149] Nikitin, A. M.; Lyubartsev, A. P. *J. Comput. Chem.* **2007**, *28*, 2020.

[150] Jellema, R.; Bulthuis, J.; v. d. Zwan, G. *J. Mol. Liq.* **1997**, *73*, 179.

[151] Barthel, J.; Gores, H. J.; Schmeer, G.; Wachter, R. In *Topics in Current Chemistry*; Boschke, F. L., Ed.; Springer: Berlin/Heidelberg, 1983; Vol. 111, pp 33–144.

[152] Ohba, T.; Ikawa, S. *Mol. Phys.* **1991**, *73*, 999.

[153] Hirata, Y. *J. Phys. Chem. A* **2002**, *106*, 2187.

[154] Firman, P.; Marchetti, A.; Xu, M.; Eyring, E. M.; Petrucci, S. *J. Chem. Phys.* **1991**, *95*, 7055.

[155] Ohba, T.; Ikawa, S.; Kimura, M. *Chem. Phys. Lett.* **1985**, *117*, 397.

[156] Arnold, K. E.; Yarwood, J.; Price, A. H. *Mol. Phys.* **1983**, *48*, 451.

[157] Yarwood, J. *J. Mol. Liq.* **1985**, *31*, 79.

[158] Yarwood, J. *J. Mol. Liq.* **1987**, *36*, 237.

[159] Vij, J. K. *Int. J. Infrared and Millimeter Waves* **1989**, *10*, 847.

[160] Lyashchenko, A. K.; Novskova, T. A. *Zh. Fiz. Khim.* **2002**, *76*, 1949.

[161] Hunger, J.; Stoppa, A.; Thoman, A.; Walther, M.; Buchner, R. *Chem. Phys. Lett.* **2009**, *471*, 85.

[162] Venables, D. S.; Chiu, A.; Schmuttenmaer, C. A. *J. Chem. Phys.* **2000**, *113*, 3243.

[163] Park, S.; Flanders, B. N.; Shang, X.; Westervelt, R. A.; Kim, J.; Scherer, N. F. *J. Chem. Phys.* **2003**, *118*, 3917.

[164] Perng, B.; Ladanyi, B. M. *J. Chem. Phys.* **1999**, *110*, 6389.

[165] Dolores Elola, M.; Ladanyi, B. M. *J. Chem. Phys.* **2005**, *122*, 224506.

[166] Fukasawa, T.; Sato, T.; Watanabe, J.; Hama, Y.; Kunz, W.; Buchner, R. *Phys. Rev. Lett.* **2005**, *95*, 197802.

[167] Vij, J. K.; Reid, C. J.; Evans, M. W. *Chem. Phys. Lett.* **1982**, *92*, 528.

[168] Woods, K. N.; Wiedemann, H. *J. Chem. Phys.* **2005**, *123*, 134506.

[169] Yomogida, Y.; Sato, Y.; Nozaki, R.; Mishina, T.; Nakahara, J. *J. Mol. Liq.* **2010**, *154*, 31.

[170] Hasted, J. B. *Aqueous Dielectrics*; Chapman & Hall: London, UK, 1973.

[171] Heintz, A.; Klasen, D.; Lehmann, J. K.; Wertz, C. *J. Solution Chem.* **2005**, *34*, 1135.

[172] Martins, R. J.; de M. Cardoso, M. J. E.; Barcia, O. E. *Ind. Eng. Chem. Res.* **2000**, *39*, 849.

[173] Kelkar, M. S.; Maginn, E. J. *J. Phys. Chem. B* **2007**, *111*, 4867.

[174] Bingham, E. C. *Fluidity and Plasticity*; McGraw-Hill: New York, USA, 1922.

[175] Grunberg, L.; Nissan, A. H. *Nature* **1949**, *164*, 799.

[176] Stewart, J. J. P. MOPAC2009, Stewart Computational Chemistry, Colorado Springs, CO, USA.

[177] Weingärtner, H.; Sasisanker, P.; Daguenet, C.; Dyson, P. J.; Krossing, I.; Olleinikova, A.; Slattery, J.; Schubert, T. *J. Phys. Chem. B* **2007**, *111*, 4775.

[178] Daguenet, C.; Dyson, P. J.; Krossing, I.; Olleinikova, A.; Slattery, J.; Wakai, C.; Weingärtner, H. *J. Phys. Chem. B* **2006**, *110*, 12682.

[179] de Andrade, J.; Böes, E. S.; Stassen, H. *J. Phys. Chem. B* **2008**, *112*, 8966.

[180] Schröder, C.; Steinhauser, O. *J. Chem. Phys.* **2008**, *128*, 224503.

[181] Schröder, C.; Haberler, M.; Steinhauser, O. *J. Chem. Phys.* **2008**, *128*, 134501.

[182] *NIST Ionic Liquids Database, ILThermo. NIST Standard Reference Database 147*. National Institute of Standards and Technology, Standard Reference Data Program: Gaithersburg, MD, 2006. http://ILThermo.boulder.nist.gov/ILThermo/.

[183] Woodward, C. E.; Harris, K. R. *Phys. Chem. Chem. Phys.* **2010**, *12*, 1172.

[184] Onsager, L.; Fuoss, R. M. *J. Phys. Chem.* **1932**, *36*, 2689.

[185] Barthel, J. M. G.; Krienke, H.; Kunz, W. *Physical Chemistry of Electrolyte Solutions—Modern Aspects*; Steinkopf, Springer: Darmstadt, New York, 1998.

[186] Chirico, R. D.; Diky, V.; Magee, J. W.; Frenkel, M.; Marsh, K. N.; Rossi, M. J.; McQuillan, A. J.; Lynden-Bell, R. M.; Brett, C. M. A.; Dymond, J. H.; Goldbeter, A.; Hou, J.-G.; Marquardt, R.; Sykes, B. D.; Yamanouchi, K. *Pure Appl. Chem.* **2009**, *81*, 791.

[187] Wang, J.; Tian, Y.; Zhao, Y.; Zhuo, K. *Green Chem.* **2003**, *5*, 618.

[188] Zafarani-Moattar, M. T.; Shekaari, H. *J. Chem. Thermodyn.* **2006**, *38*, 1377.

[189] Li, W.; Zhang, Z.; Han, B.; Hu, S.; Xie, Y.; Yang, G. *J. Phys. Chem. B* **2007**, *111*, 6452.

[190] Huo, Y.; Xia, S.; Ma, P. *J. Chem. Eng. Data* **2007**, *52*, 2077.

[191] Shekaari, H.; Zafarani-Moattar, M. T. *Int. J. Thermophys.* **2008**, *29*, 534.

[192] Stoppa, A.; Buchner, R.; Hefter, G. *J. Mol. Liq.* **2010**, *153*, 46.

[193] Huddleston, J. G.; Visser, A. E.; Reichert, W. M.; Willauer, H. D.; Broker, G. A.; Rogers, R. D. *Green Chem.* **2001**, *3*, 156.

[194] Zafarani-Moattar, M. T.; Majdan-Cegincara, R. *J. Chem. Eng. Data* **2007**, *52*, 2359.

[195] Gutman, V. *The Donor–Acceptor Approach to Molecular Interactions*; Plenum: New York, 1978.

[196] Barthel, J.; Deser, R. *J. Solution Chem.* **1994**, *23*, 1133.

[197] Wismeth, E. Ph.D. thesis, Regensburg, 1996.

[198] Barthel, J.; Kleebauer, M. *J. Solution Chem.* **1991**, *20*, 977.

[199] Mandal, P.; Sarkar, M.; Samanta, A. *J. Phys. Chem. A* **2004**, *108*, 9048.

[200] Paul, A.; Mandal, P. K.; Samanta, A. *J. Phys. Chem. B* **2005**, *109*, 9148.

[201] Wang, Y. T.; Voth, G. A. *J. Am. Chem. Soc.* **2005**, *127*, 12192.

[202] Schröder, C.; Rudas, T.; Steinhauser, O. *J. Chem. Phys.* **2006**, *125*, 244506.

[203] Habasaki, J.; Ngai, K. L. *J. Chem. Phys.* **2008**, *129*, 194501.

[204] Katoh, R. *Chem. Lett.* **2007**, *36*, 1256.

[205] Jin, H.; O'Hare, B.; Dong, J.; Arzhantsev, S.; Baker, G. A.; Wishart, J. F.; Benesi, A. J.; Maroncelli, M. *J. Phys. Chem. B* **2008**, *112*, 81.

[206] Mele, A.; Tran, C. D.; De Paoli Lacerda, S. H. *Angew. Chem. Int. Ed.* **2003**, *42*, 4364.

[207] Katayanagi, H.; Nishikawa, K.; Shimozaki, H.; Miki, K.; Westh, P.; Koga, Y. *J. Phys. Chem. B* **2004**, *108*, 19451.

[208] Semiempirical calculations of various confomers have been performed using MOPAC2009[176] and the PM6 Hamiltonian. Dipole moments were calculated assuming the geometric center as the pivot. Molecular diameters were obtained by taking the longest distance between two atoms and adding the van der Waals radii of the atoms.[216] These diameters were used to calculate the maximum molecular volumes of reorientation. To account for solvent effects the COSMO[217] technique was applied. Winmostar[218] was used to determine van der Waals volumes from the optimized geometry.

[209] Katsuta, S.; Imai, K.; Kudo, Y.; Takeda, Y.; Seki, H.; Nakakoshi, M. *J. Chem. Eng. Data* **2008**, *53*, 1528.

[210] Barthel, J.; Iberl, L.; Rossmaier, J.; Gores, H. J.; Kaukal, B. *J. Solution Chem.* **1990**, *19*, 321.

[211] Wang, H.; Wang, J.; Zhang, S.; Pei, Y.; Zhuo, K. *ChemPhysChem* **2009**, *10*, 2516.

[212] Rowlen, K. L.; Harris, J. M. *Anal. Chem.* **1991**, *63*, 964.

[213] Kiefer, J.; Fries, J.; Leipertz, A. *Appl. Spectrosc.* **2007**, *61*, 1306.

[214] Kabisch, G. *Z. Phys. Chem.* **1982**, *263*, 48.

[215] Barthel, J.; Buchner, R.; Wismeth, E. *J. Solution Chem.* **2000**, *29*, 937.

[216] Bondi, A. *J. Phys. Chem.* **1964**, *68*, 441.

[217] Klamt, A.; Schüürmann, G. *J. Chem. Soc. Perkin Trans.* **1993**, *2*, 799.

[218] Senda, N. Winmostar, version 3.78f. http://winmostar.com.

I want morebooks!

Buy your books fast and straightforward online - at one of world's fastest growing online book stores! Environmentally sound due to Print-on-Demand technologies.

Buy your books online at
www.morebooks.shop

Kaufen Sie Ihre Bücher schnell und unkompliziert online – auf einer der am schnellsten wachsenden Buchhandelsplattformen weltweit! Dank Print-On-Demand umwelt- und ressourcenschonend produziert.

Bücher schneller online kaufen
www.morebooks.shop

KS OmniScriptum Publishing
Brivibas gatve 197
LV-1039 Riga, Latvia
Telefax: +371 686 204 55

info@omniscriptum.com
www.omniscriptum.com

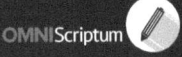

Printed by Books on Demand GmbH, Norderstedt / Germany